Perennials for the Landscape

Perennials for

Dianne A. Noland

Instructor
University of Illinois

Kirsten Bolin

Plant Buyer
Van Zelst, Inc.

the Landscape

Interstate Publishers, Inc.
Danville, Illinois

Perennials for the Landscape

Some computer based art by
Electronic Illustrators Group

Library of Congress Catalog Number 99-71696

ISBN 08134-3149-2

1 2 3 4 5 6 7 8 9 10   04 03 02 01 00 99

Order from

Interstate Publishers, Inc.

510 North Vermilion Street
P.O. Box 50
Danville, IL 61834-0050

Phone: (800) 843-4774
Fax: (217) 446-9706
Email: info-ipp@IPPINC.com

World Wide Web: http://www.IPPINC.com

Introduction

What is a flower? In a technical sense, it is a means for plant reproduction by producing a fruit containing a seed for the next generation. At the same time, it can be a source of food for birds and animals. But to humans, it is much more than one of nature's creations. Flowers represent emotional symbols of hope, cheer, love, and even evoke memories of times past. The physical nature of the flower appeals not only to the emotional but to the physical senses as well. The vibrant colors, unique shapes, varied textures, and sweet fragrances of flowers awaken the senses of sight, touch, and smell. I have been drawn to flowers, almost irresistibly, since the age of five because of all of these factors.

Annuals, biennials, herbaceous perennials, and woody perennials all produce flowers of some kind. This book focuses only on the herbaceous perennials, plants that have been a life-long love of mine since I began to garden. The herbaceous perennials included in this book are a good representation of the most common, reliable, and beautiful ones, which I found to be the most appropriate to teach my university students.

A *herbaceous perennial* is defined as a non-woody plant that has a life span of more than two years and completes a vegetative and reproductive phase annually. Perennials are either evergreen, that is, they will not drop their leaves but remain green, or they will die back and overwinter below the ground. These perennials overwinter as an underground structure, such as a bulb, crown, rhizome, tuber, or corm. In the subsequent pages, the term *perennial* will include only those that are herbaceous.

This book is organized so it can be used as a quick reference guide or as an identification source. As a reference guide, each plant is thoroughly described, from the flowers and leaves to the landscape use and growing requirements. In using this book as an identification source, the categorizations will allow the reader to identify plants by season of bloom and where the plants are placed within the landscape.

Each plant is listed alphabetically by the scientific name according to the plant's flowering season. The Royal Horticultural Society's reference books—*Dictionary of Gardening* (1992) and *Index of Garden Plants* (1994)—were very helpful in labeling these plants with the most current and correct names.

The season of flowering category is then subdivided into herbaceous perennial plants and bulbs. *Season of flowering* is the time span in which the plant produces its flower. Most plants are grown for their flowers, thus, the categorization into flowering times. There are four categories: early spring, spring, summer, and mid to late summer and fall as follows:

> <u>**Early Spring**</u> flowering plants may start to flower as early as late February and or any time through late April.

> <u>**Spring**</u> includes those plants flowering in May and June.

> <u>**Summer**</u> covers plants blooming in June, July, and possibly into August.

> <u>**Mid to Late Summer to Fall**</u> covers all the plants that flower from mid to late July through August until frost.

The next category for identification purposes is the *landscape use*, or more simply, where a plant is placed within the landscape. Shorter, smaller plants should be placed in front of taller, larger plants. This book divides plants into five groupings: ground cover, edging, foreground, midborder and background plants. *Ground cover* includes plants that spread laterally and cover the ground. The height in the ground cover category is usually not more than 12 inches tall. The plants that fall into the *edging* category are usually those that are 6 to 12 inches. The difference between edging plants and ground covers is an edging will not spread as generously but tends to mound at the edge of the bed. *Foreground* plants are 1 to 2 feet in height and placed just behind the edging plant. *Midborder* plants are usually taller, 2 to 3 feet tall. Lastly, *background* plants serve as a backdrop for smaller plants. These are the tallest plants in the border—3 feet and up. Of course, a plant's placement is relevant to the size of the garden, but, in general, these groupings by height make for a good identification feature.

Once a plant is identified by these categories, it is then described by the following categories: flower characteristics, leaf morphology, growing needs of the plant, landscape use of the plant, design ideas, popular cultivars and related species, problems that may arise, propagation, and hardiness zones.

The first trait of the plant to be described is the most obvious, the flower. Each inflorescence and flower is described by type, size, color, and any other characteristics that make it unique.

Next, the leaves are described by morphology, leaf arrangement, size, color, and texture. Leaves may arise from the underground growth structure, i.e., basal leaves, or be arranged on the stem in either an alternate, opposite, or whorled fashion.

To grow a plant, the proper growing conditions and requirements must be known. This information is included under the heading of "Growing needs."

Light requirements and soil conditions are listed here. The light requirements are divided into sun, partial shade, and shade (full or dense shade).

○ — a sunny location — has less than two to four hours of shade in an eight-hour day. Because of the intense heat, most sunny, western exposures are considered full sun even though the site is shady in the morning.

◑ — a partially shaded site — has at least four to six hours of shade within the day. Dappled sunlight throughout the entire day could also be considered a partially shaded site.

● — a shady location (full or dense shade) — has filtered sunlight for two hours or less throughout the day.

Information about preferred soil conditions, such as soil moisture, fertilization, pH, soil amendments, and any mulching requirements, are also included within this section.

In general, an excellent site must be accompanied by preparing the bed well. Lay out the shape of the bed using stakes and string or a garden hose (like I use) for curved beds. Remove any sod. Work the soil to the depth of the tiller or double dig the bed for deep-rooted perennials. Add compost or peat moss as needed, depending upon the clay content.

Once one knows how to grow the plant, one must find the proper place within the landscape in which to place the plant. The "Landscape use" heading identifies the height or size, habit or shape, texture ("feel" or "look" of the plant), and recommended spacing requirements. Effective uses and placement in the garden border will be discussed for each plant. After establishing a plant's place in the garden, the "Design ideas" section will give examples of companion plants. The design ideas are only suggestions to use in beginning a garden.

Following these headings is a section listing recommended and outstanding "Cultivars." A *cultivar* is a cultivated variety, bred and grown to show off specific characteristics. Notable related species may also be mentioned including a brief description.

Under the heading of "Problems" is a listing of any serious insect or disease problems that would greatly affect the appearance or longevity of the plant.

The "Propagation" section tells how to start new plants. Propagation is generally carried out as division, seeds, or cuttings.

Division of the plant is the actual lifting of the entire plant and gently dividing it into smaller sections. Each division can then be replanted in the garden.

Seeds can be purchased or collected and sown indoors in early spring. Maintain high humidity and even moisture during germination. Seedlings can be transplanted into a larger container and then planted outdoors in the late spring. A few of the good self-sowers (purple coneflower — *Echinacea* or blanketflower — *Gaillardia)* can be directly sown into the garden. Most perennial plants do not flower the first year from seed. A few exceptions include Shasta daisies, some yarrows, and black-eyed Susans.

Cuttings (stem or root) must be taken to produce a cultivar. Stem cuttings usually require rooting hormone, will root in two to six weeks in constant high humidity and moisture in a porous soil, and should be transplanted to a container for further root establishment before planting in the garden. Cuttings will reward you with identical plants (clones) to the parent.

The last heading is "Zones." This heading gives the range in which the plant is hardy. The hardiness of a plant is its ability to survive in a particular environment. Each climatic region is given a rating or zone. Match your zone to the ranges given to determine hardiness in your area.

The following pages will help the novice or the professional develop a beautiful perennial garden, one that will bring enjoyment and memories for years to come. Whether a beginner or a long-time gardener, one can spend minutes or hours a day enjoying and working with nature. The information contained in this book will help expand your ideas and allow your garden and your ideas to bloom.

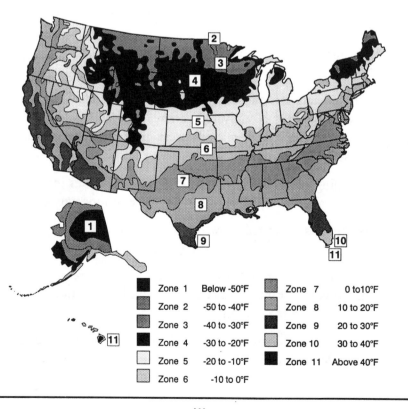

	Zone 1	Below -50°F		Zone 7	0 to 10°F
	Zone 2	-50 to -40°F		Zone 8	10 to 20°F
	Zone 3	-40 to -30°F		Zone 9	20 to 30°F
	Zone 4	-30 to -20°F		Zone 10	30 to 40°F
	Zone 5	-20 to -10°F		Zone 11	Above 40°F
	Zone 6	-10 to 0°F			

Sample Entry Page

Scientific Name (Old Botanical Name)

Common name
Family
Sun or Shade, Landscape use, Season of bloom

Flower description

Leaf description

Growing needs

Use within the Landscape

Design ideas

Popular cultivars (and Related Species when applicable)

Serious problems (Insects and diseases)

Propagation methods

Zone ranges

Acknowledgments

A big thank you—

- To my parents Neil and Erma who first spotted my gardening disease (to my dad for committing many memorable horticultural faux pas, which I pass on to educate and entertain my gardening audiences and classes)

- To my flower gardening grandmothers — Lola Moma, Margaret Scott, and Ruth Noland and vegetable gardening grandfathers — Charley Noland and James Moma whose genetic make-up led to my gardening fever

- To my perennials students at the University of Illinois since 1984 who have helped me to be a better teacher

- To my perennials teacher Don Saupe for his knowledge and the use of some of the slides from his huge collection

- To colleagues and gardening friends especially Sandra Casserly who have shared plants, knowledge, and gardening enthusiasm

- To my husband Dave Negangard and my son Drew for all their support, interest, and patience as I talk about plants all the time

Dianne Noland

I want to express my appreciation to Dianne Noland and Art Spomer at the University of Illinois for helping me on this project. Thanks also to Ken Robertson and Bill Sullivan for their advice. This project was a definite learning experience. I am also grateful to my parents for helping me get where I am today.

Kirsten Bolin

About the Authors

Dianne A. Noland

Dianne Noland is a University of Illinois horticulture instructor of six courses, covering identification and landscaping with perennials, herbs, and annuals, as well as floral design topics. She is an enthusiastic teacher of nearly 300 students per year. Mrs. Noland has been awarded numerous teaching honors by her peers and students alike over her 20 years as an educator, including a 1985 Teaching Award of Merit from the National Association of Colleges and Teachers of Agriculture, a 1990 Sustained Excellence Award from the University of Illinois College of Agriculture, and a 1996 induction into the Academy of Teaching Excellence for the College of Agricultural, Consumer, and Environmental Sciences at the University of Illinois. She has been chosen by her students as an Outstanding Instructor every semester since she began teaching as a faculty member in 1980.

Gardening became an interest and a passion when Dianne was a child. Her interest was further fueled by an internship with a horticultural family in Switzerland. She earned two horticulture degrees (B.S. in Ornamental Horticulture and M.S. in Horticulture from the University of Illinois) and gained valuable work experience floral designing and working at three nurseries (one with a garden center) before accepting the teaching position at the University of Illinois.

Enjoying every facet of flowers and their design, she has also co-authored seven floral design books, including *The Centennial History of Floral Design*, published by Florists' Review and six volumes of *Encyclopedia of Floristry*, published by The Redbook Floral Services.

Dianne Noland often shares her garden, located on 40 acres along the Salt Fork River, through video presentations as a regular on Illinois Gardener, a PBS TV show in central Illinois. Prior to that, Dianne was a regular for 10 years on Garden Talk, a University of Illinois gardening segment on the local television station. She and her husband Dave have hosted four flower and garden tours to Europe. She is a popular speaker throughout Illinois on gardening and landscaping with flowers.

Kirsten Bolin

Kirsten Bolin is a plant buyer with a Chicago area landscaping company. She works daily with perennials and their use in the landscape. Through regular travel, she is exposed to new perennial varieties and receives feedback from peers on the use of perennials in the landscape.

Ms. Bolin holds two degrees from the University of Illinois. It was while at the University of Illinois that she became first a student, then a friend of her coauthor. They have collaborated extensively and share a passion for perennials and their use in the landscape. Bolin's work on her master's thesis provided the foundation for much of the information contained in this book.

Contents

*L*ist of Perennials by Botanical Name

List of Perennials by Common Name

EARLY SPRING PERENNIALS

Arabis caucasica

Rockcress, Wall Rockcress
Brassicaceae (Mustard Family)
○, ◑ Edging, Early spring

Flowers: The flowers of rockcress are white or blush pink and delicately fragrant. They are four-petalled, ½-inch across, and arranged in loose racemes above the foliage.

Leaves: The gray-green leaves are usually evergreen, making rockcress an important year-round landscape feature. Each 1-inch long leaf is ovate with toothed wavy margins and soft stellate pubescence and is alternate on the stem.

Growing Needs: Rockcress prefers a full sun, well-drained site. It is drought tolerant and will not tolerate wet soils. Prune back after flowering to promote compact growth. Site in light shade where summers are hot.

Landscape Use:

Height:	6 to 12 inches
Habit:	Low mounded
Spacing:	12 to 15 inches
Texture:	Medium

Rockcress is an excellent, highly floriferous edging plant in the early sunny spring border. It looks equally beautiful draping over walls or in the rock garden.

Design Ideas: Plant rockcress with purple dwarf iris for an outstanding colorful duet in early spring — quite a show stopper! Combine rockcress with the early bulbs that lose their foliage in summer, such as glory of the snow and crocus.

Cultivars:

'Flore Pleno' is a long-flowered double, 8 inches tall.
'Snow Cap' is white-flowered, 6 to 8 inches.
'Variegata' has creamy white variegated leaves.
'Spring Charm' is light pink.

Problems: None serious.

Propagation: Division, cutting, or seed.

Zones: 4-7

Related Species:

Arabis procurrens, creeping rockcress, has white flowers, forms a spreading mat of shiny evergreen leaves, and also prefers full sun, well-drained soil.

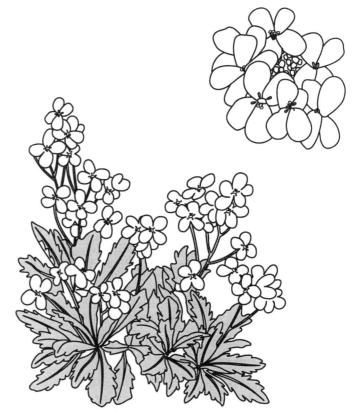

Aubrieta deltoidea

Purple Rockcress
Brassicaceae (Mustard Family)
○, ◑   Edging, Early spring

Flowers: Small, ¾-inch, purple or pink, four-petalled flowers are arranged in loose racemes. In spring, this plant is covered with flowers.

Leaves: These alternate, small, 1-inch, attractive gray-green obovate to spatulate leaves have rounded, wavy margins. Fuzzy stellate pubescence covers each leaf. This plant has leaves similar to, but smaller than, rockcress.

Growing Needs: Purple rockcress does well in partial shade to full sun. Soils should be light and porous and, of course, well drained. For best results, plant and divide in spring. This plant is drought and lime tolerant. To maintain compact growth, shear after flowering.

Landscape Use:

Height:	4 to 8 inches
Habit:	Low mounded
Spacing:	12 inches
Texture:	Medium

Purple rockcress looks great as an edging, in a rock garden, or as a container plant. It also is a charmer when cascading over low walls.

Design Ideas: This low-grower looks great with any spring bulbs, as well as rockcress and basket of gold. Purple rockcress used en masse displays subtle variations in color from purples to pinks and is a pleasing scene, especially when used as an edging with the yellowish cushion spurge or tulips (try Shirley tulips — white with lavender edges).

Cultivars:

'Aureovariegata' has yellow-edged leaves and light purple flowers.

'Borsch's Flower' has a white flower.

'Cambellii' has double, lavender-blue flowers.

'Greencourt Purple' is a double flower.

'Purple Cascade' is compact, 4 inches.

'Red Cascade' has reddish flowers, 4 to 6 inches tall.

Problems: None serious.

Propagation: Division in spring, cuttings, or by layering.

Zones: 4-8 (Outstanding in cooler climates.)

Aurinia saxatilis

(Alyssum saxatile)

Basket of Gold
Brassicaceae (Mustard Family)
○ Edging, Early spring

Flowers: The four-petalled, notched, ¼-inch flowers range from lemon yellow to golden yellow and are arranged in corymbose panicles. Individual flowers are tiny but when viewed as a whole, make quite a showy spring spectacle.

Leaves: The woolly gray oblanceolate leaves of this plant are alternate and 2 to 5 inches long — large compared to rockcress and purple rockcress. Its fuzzy leaves are a nice textural contrast to the flower and remain attractive throughout the growing season.

Growing Needs: Basket of gold prefers a full sun, well-drained site. Too much moisture or fertility can cause sprawling, rotting, or death of the plant. Pruning after flowering will produce a more compact plant and possibly a rebloom.

Landscape Use:

Height:	6 to 12 inches
Habit:	Mounded
Spacing:	12 to 18 inches
Texture:	Medium

Basket of gold is a natural as a sunny edging plant or a draping plant to soften architecture (sidewalks or walls). It is often used in rock gardens.

Design Ideas: Basket of gold masses beautifully with clumps of red mid-season tulips, purple pansies, and purple rockcress. As the sole planting in the rock crevices of retaining walls, it can make an impressive blanket.

Cultivars:

'Argentea' has very silvery leaves.

'Citrinum' or 'Silver Queen' are both 10 inches with lemon yellow flowers.

'Compacta' is 6 to 8 inches tall.

'Dudley Neville' has light-orange flowers.

'Dudley Neville Variegated' has white variegated leaves.

'Sulphureum' is a compact 8 to 10 inches with striking yellow flowers.

'Sunny Border Apricot' is 8 inches with apricot-colored flowers.

'Variegata' has light-green leaves with white variegation, weak growing.

Problems: None serious.

Propagation: Division, seed, and by cuttings.

Zones: 3-7

Bergenia cordifolia

(Saxifraga cordifolia)

Heartleaf Bergenia
Saxifragaceae (Saxifrage Family)
◐, ○ Foreground, Early spring

Flowers: Pink or white bell-shaped flowers (¼- to ½-inch) are borne in 6- to 8-inch panicles on bold, thick scapes just above the foliage. The flowers are beautiful, but often considered second in importance to the foliage.

Leaves: The bold, rounded leaves are large, fleshy, and glossy ranging in size from 10 to 12 inches long and 6 to 8 inches wide with long, thick petioles. The leaves are evergreen, often displaying a reddish to purplish fall/winter coloring. The newly emerging leaves may appear reddish and are alternate on the stem.

Growing Needs: This is a dependable, long-lasting plant for a moist, moderately fertile soil with partial shade. However, this plant will tolerate sun, dense shade (fewer flowers), and drought.

Landscape Use:

Height:	12 to 18 inches
Habit:	Bold clusters
Spacing:	18 inches
Texture:	Bold

At maturity, heartleaf bergenia forms extremely striking accent plants in the partially shady garden. This plant is an effective ground cover, especially attractive near water. Patience is required when establishing heartleaf bergenia since immature plantings may appear more like a garden of cabbage plants.

Design Ideas: Effective uses of heartleaf bergenia are along woodland edges with drifts of spring bulbs and ferns nearby. Another pleasing combination combines European wild ginger as an edging and shrubs, such as rhododendrons, behind it.

Cultivars:

'Baby Doll' is pale pink, 12 inches.

'Bressingham White' has white flowers and robust foliage, 12 inches.

'Morning Red' has dark purplish red flowers and bronze leaves, 12 inches.

'Perfect' has dark red flowers and purple winter foliage, 10 to 12 inches.

'Purpurea' has red flowers, reddish stems, and good reddish purple fall and winter color.

'Rotblum' has red flowers and bronze winter foliage, 10 inches.

Problems: None serious.

Propagation: Division in spring and seed.

Zones: 3-8

Euphorbia polychroma *(epithymoides)*

Cushion Euphorbia or Spurge Euphorbiaceae
(Spurge or Poinsettia Family)
○   Foreground, Early spring

Flowers: This plant is covered with cheerful bright yellow to greenish yellow flowers in the spring. The ¾-inch flowers are actually petaloid bracts (like poinsettias) and are borne in an umbel-like cyme called a cyathium. The true flowers are greenish and unnoticeable.

Leaves: The 2-inch leaves of cushion spurge are alternate (appearing whorled), dark green, and oblong, often with a noticeable midvein. This plant exudes milky sap when a leaf is removed (making it easy to identify). The foliage is attractive all season long and exhibits a dark red fall color.

Growing Needs: Cushion spurge performs well in full sun, well-drained sites. This long-lived plant can take drought conditions but does not transplant easily.

Landscape Use:

Height:	12 to 18 inches
Habit:	Symmetrical mound
Spacing:	18 to 24 inches
Texture:	Medium

Cushion euphorbia is a very showy plant for the sunny foreground. It makes an outstanding specimen plant.

It is not recommended for massed use since each very mounded plant does not gracefully blend with its neighbor, forming a "bumpy" landscape.

Design Ideas: A spring delight is cushion euphorbia with grape hyacinths and evergreen candytuft. Combine it with primroses and spring bulbs (hides their faded summer foliage, too) or with sea pink (*Armeria*) and creeping phlox. In late summer, cushion spurge's leaf color and texture look fantastic with blue plumbago (*Ceratostigma*) in flower.

Cultivars:

'Emerald Jade' has good fall color.
'Midas' has bright yellow flowers.
'Major' is very mounded with bright yellow flowers.
'Purpurea' has yellow flowers and purple leaves.

Problems: None serious.

Propagation: Seed and careful spring division of young plants.

Zones: 3-9

Euphorbia myrsinites

Myrtle Euphorbia or Spurge Euphorbiaceae
(Spurge Family)
○   Edging, Early spring

Flowers: The 2-inch flower head is an umbel-like cyathium of petaloid bracts. The bract color is yellow to chartreuse green. Each petaloid bract is cup-shaped.

Leaves: The ½-inch, obovate, blue-green, succulent leaves are whorled on the stem and exude a milky sap when broken. This plant is grown for its interesting foliage and is often evergreen.

Growing Needs: Myrtle spurge is a drought tolerant plant that does well in a sunny, well-drained site. It does not transplant well. Trimming back faded "flowers" will lessen the excess of volunteer seedlings. The seeds are dehiscent and, when mature, are propelled everywhere with great force leading to many new seedlings.

Landscape Use:

Height:	6 to 9 inches
Habit:	Prostrate and sprawling
Spacing:	18 to 24 inches
Texture:	Coarse in flower, medium in foliage.

Myrtle euphorbia makes an excellent ornamental foliage in the sunny garden. Use this plant as a limited ground cover or in a rock garden setting. Place it in tough spots where it is hard to establish other plants.

Design Ideas: Myrtle spurge looks nice with creeping phlox and sedum. Use it to cascade over a stone wall or in the rock garden.

Cultivars:

'Washfield' has reddish foliage.

Problems: None serious.

Propagation: Seed (collect it yourself) and careful transplanting of small seedlings.

Zones: 5-10

Helleborus niger

Christmas Rose
Ranunculaceae (Buttercup Family)
◑, ● Edging, Foreground, Ground cover,
Early spring

Flowers: This early flowering plant produces 2- to 3-inch, creamy white solitary flowers, each with five showy sepals, on a leafless peduncle, or flower stalk. The flowers pinken with age or due to cold.

Leaves: The 6- to 8-inch, dark green leaves are palmately divided into pointed ovate segments. Each basal leaf is long petioled and has serrated margins. The foliage often appears after flowering. In milder regions, the foliage will remain evergreen.

Growing Needs: Christmas rose performs best in a protected, partially shaded site with a moist, slightly acidic, well-drained soil. A light mulch and moderate fertilization is beneficial.

Landscape Use:

Height:	12 inches
Habit:	Mounded
Spacing:	12 inches
Texture:	Medium bold

Christmas rose makes an excellent addition to the shade border. It makes a good limited ground cover and is an excellent cut flower and foliage.

Design Ideas: Group this plant with Japanese painted fern and wildflowers, such as wild larkspur, Virginia bluebells, and *Trillium* species. Early bulbs, such as snowdrops and winter aconite, combine well together. Christmas rose also make a nice edging plant for a shady shrub border.

Cultivars:

'Potters Wheel' has a nice white flower with a green "eye" or center.

Problems: Slugs and crown rot.

Propagation: Seed or spring division.

Zones: 3-9

Related Species:

Helleborus orientalis, Lenten rose is similar to Christmas rose but flowers later, is slightly taller (12 to 18 inches), and is less cold hardy. There is also a greater range of flower colors in lenten rose as follows:

> *H. orientalis* ssp. *orientalis* has pure white to cream flowers.

> *H. orientalis* ssp. *abchasicus* has reddish purple flowers, finely spotted with darker red or purple.

> *H. orientalis* ssp. *guttatus* is white or cream with distinct darker reddish purple spotting or blotches.

Helleborus purpurascens has dark purple, reddish, purple, green, or brown flowers, often darker on the outside, 12 to 18 inches.

'Red Power' has maroon red flowers.

Iberis sempervirens

Evergreen Candytuft
Brassicaceae (Mustard Family)
○   Edging, Early spring

Flowers: The flowers of candytuft are white, ¼-inch and borne in wide 1- to 1½-inch rounded racemes with flowers clustered at the top. Each flower has two shorter and two longer petals. In spring, this plant is literally smothered with white flowers. Evergreen candytuft usually flowers slightly later than the other Brassicaceae plants (*Arabis, Aubrieta* and *Aurinia* listed earlier).

Leaves: The attractive alternate leaves are small, 1 to 2 inches long, narrowly oblong and shiny dark green. The foliage remains nice all through the summer and fall and is usually semi-evergreen.

Growing Needs: Evergreen candytuft prefers a full sun location. The soil should be well drained and moderately fertile. The plant may be pruned back after flowering for a more compact plant.

Landscape Use:

    Height: 6 to 12 inches
    Habit: Mounded
    Spacing: 12 to 18 inches
    Texture: Medium

Evergreen candytuft is a great edging plant or ground cover. This prolific flowerer remains attractive after flowering due to its foliage.

Design Ideas: Red tulips combine well with evergreen candytuft for early color. For late spring and early summer color, combine it with coral bells (*Heuchera*) and pinks (*Dianthus*). Pasque flower (*Pulsatilla vulgaris*) provides a breathtaking combination with evergreen candytuft.

Cultivars:

'Autumn Snow' and 'Autumn Beauty' have a small rebloom in the fall, 8 to 10 inches.
'Compacta' has a compact habit, 6 inches.
'Little Gem' is a compact 6 inches.
'Purity' is compact, 8 inches, and covered with white flowers in spring.
'Snowflake' has large flowers and leaves and grows to 10 inches.

Problems: None serious.

Propagation: Seed, careful spring division, and by summer cuttings.

Zones: 3-9

Phlox subulata

Moss Phlox, Creeping Phlox
Polemoniaceae (Phlox Family)
○ Edging, Early spring

Flowers: The flower of moss phlox is five-petalled; each petal is shallowly notched. These ½-inch flowers come in an array of colors—from pink to white, lavender, red-purple to bluish-purple. The flowers may be a solid color or have a contrasting "eye" (center). The flowers are borne in a terminal cyme.

Leaves: The wiry stems are crowded with stiff, opposite to whorled, 1-inch evergreen leaves. Each leaf is linear and marginally hairy. The leaves appear prickly and needle-like.

Growing Needs: Moss phlox does best in a well-drained, sunny location. To promote denser foliage, shear back any straggly growth following flowering. There may also be a sparse rebloom.

Landscape Use:

Height:	4 to 8 inches
Habit:	Prostrate to low mounded
Spacing:	18 inches
Texture:	Medium fine

Moss phlox makes a beautiful spring edging plant. It also delightfully drapes over walls and is excellent for slopes and banks. It is most appealing when massed as one color.

Design Ideas: White or pink moss phlox combines well with grape hyacinths and tulips. Lavender moss phlox looks showy when combined with *Ajuga* 'Burgundy Glow', lamb's ear, and *Iris pallida* 'Argentea Variegata'.

Cultivars:

'Blue Emerald' has blue flowers.
'Candy Stripe' has pink and white striped flowers.
'Fort Hill' is pink flowered and reblooms.

'Scarlet Flame' has a bright rosy-red flower with a darker eye.
'White Delight' has a pure white flower.

Problems: None serious. Occasionally, spider mites during hot, dry summers.

Propagation: Layering, division after flowering, and cuttings.

Zones: 3-9

Related Species and Hybrids:

Phlox divaricata, wild sweet William or wild blue phlox, is a lovely semi-evergreen woodland flower with blue to lavender clusters of 1-inch flowers in

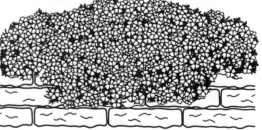

spring. The summer foliage is a great bonus, 8 to 12 inches. Sun/shade.

'Dirigo Ice' is pale lavender blue.

'Fuller's White' is a prolific white form.

ssp. *laphamii* has darker blue flowers, 10 inches.

'Chattahoochee' (*P. divaricata* ssp. *laphamii* × *P. pilosa*) is bluish-lavender with a deeper maroon eye, very free flowering.

'Clouds of Fragrance' fragrant lavender.

'London Grove' is prolific lavender-blue and fragrant, 10 inches.

Phlox stolonifera, creeping phlox, flowers after *P. subulata* and has attractive oval leaves that persist through the summer. It prefers partial shade and moist soil.

'Blue Ridge' has broad blue petals and glossy leaves, 12 inches.

'Bruce's White' is white with yellow centers.

'Home Fires' is deep pink.

'Pink Ridge' is the pink form of 'Blue Ridge'.

'Rosea' is light pink.

Phlox douglasii is a 6- to 8-inch, compact, ground or edging plant with very fragrant flowers in shades of pink, purple, mauve, and red, often with a contrasting eye.

'Crackerjack' is a floriferous pink compact plant.

'Iceberg' is white with a blue tinge.

'Red Admiral' has crimson red flowers.

Phlox × procumbens (P. stolonifera × P. subulata) is a low-growing, clump-forming, edging plant with bright purple flowers.

'Variegata' has creamy white, variegated leaves and pink flowers.

Primula × polyantha

Polyantha Primrose
Primulaceae (Primrose Family)
◑ Edging, Early spring

Flowers: Primroses may be single or double and are available in many colors, including white, yellow, orange, red, blue, and purple. The 1½-inch flower has a five-lobed corolla, often with an obvious yellow eye (center), and is delightfully fragrant. The flowers are borne in 3- to 5-inch rounded umbels.

Leaves: The 8- to 10-inch leaves are best described as puckered, wrinkled, light green, and oval with wavy margins. All of the leaves are in a basal rosette.

Growing Needs: A partially shaded, moist yet well-drained, slightly acidic site will best fulfill the needs of primrose. The addition of humus and fertilizer is also beneficial. Placement in a protected site and frequent division in early spring is recommended for a longer life of primrose; poor siting of primrose will cause it to be a very short-lived plant.

Landscape Use:

Height:	6 to 12 inches
Habit:	Rosette/two tiered
Spacing:	6 to 12 inches
Texture:	Medium

A beautiful use of primrose is as an edging plant under flowering spring trees. It also looks quite striking as a mass planting.

Design Ideas: Primrose combines well with wildflowers, such as wild blue phlox or wild ginger with Virginia bluebells and ferns behind them. A nice variety of plant habits would accent primrose. For example, plant it with Solomon's seals arching stems and the linear spiderworts. Bleeding hearts and bulbs, such as tulips and daffodils, are also nice combinations with polyantha primrose.

Cultivars:

'Crescendo Strain' has mixed colors, is reliably hardy in Zone 5, and grows to 6 inches.

Problems: Red spider mites in full sun and bacterial leaf spots.

Propagation: Seed and by division.

Zones: 3-8

Related Hardy Species:

Primula vulgaris tolerates more sun and heavier soils. It has fragrant yellow spring flowers and crinkled, evergreen foliage, 6 inches.

P. denticulata, drumstick primrose, has round clusters in purple, red, or white borne above the foliage, 8 to 10 inches.

P. japonica, candelabra primrose, grows to 18 inches in mixed colors.

Pulmonaria saccharata

Bethlehem Sage, Lungwort
Boraginaceae (Borage Family)
◐, ● Foreground, Early spring

Flowers: Bethlehem sage has tiny ½- to ¾-inch trumpet-shaped flowers borne in drooping clusters (cymes). The five-parted flowers often have the characteristic of opening pink and changing to sky blue or lavender, much like its wildflower cousin, Virginia bluebells.

Leaves: The leaves alone are a good reason to grow this plant. The leaves are borne in a basal rosette with smaller leaves arranged alternately along the stem. The leaf shape is linear to ovate with pointed tips. These 4- to 6-inch, dark-green leaves have bristly, sandpaper-like pubescence, often with attractive flecking of light green, white, or silver.

Growing Needs: A partially to fully shaded spot is preferred. Bethlehem sage will grow in full sun but in mid-summer the leaves brown and curl; new leaves will emerge in fall when temperatures are cooler. Other requirements for best growth: moisture, additional organic matter, and moderate fertilization.

Landscape Use:

Height:	12 inches
Habit:	Mounded, two tiered
Spacing:	12 to 15 inches
Texture:	Medium

This striking and dependable plant is an excellent specimen in the foreground of the shady perennial border. It is attractive massed as well. The distinctive flecking on the leaves makes this plant valuable even after flowering.

Design Ideas: Bethlehem sage mixes well with spring bulbs, especially daffodils, as well as primroses, hostas, and ferns. It also makes an excellent combination with foam flower (*Tiarella*). Plant sweet woodruff as an attractive edging with Bethlehem sage.

Cultivars:

'Alba' has white flowers, 9 inches.

'Mrs. Moon' has three colors, with pink buds, turning blue and fading to lavender, 12 inches.

'Pierre Pure Pink' has floriferous pink flowers with no hint of blue, slightly spotted leaves, and grows 12 inches.

'Pink Dawn' has pink flowers.

'Janet Fisk' has lavender pink flowers and highly flecked foliage, 8 inches.

'Roy Davidson' has sky blue flowers and heavily silver-flecked leaves, 12 inches.

'Smokey Blue' has light blue flowers and flecked leaves, 12 inches.

Problems:
Slugs and powdery mildew.

Propagation:
Seed and division in late summer or fall.

Zones:
3-8

Related Species:

P. longifolia 'Bertram Anderson' has narrow, spotted foliage and dark blue flowers, 12 inches.

P. rabra 'David Ward' has coral-colored flowers, mint green ruffled foliage with creamy white margins, and grows 14 inches.

P. officinalis 'Sissinghurst White' has white flowers and silvery flecked leaves. Site well in moist partial shade to avoid powdery mildew.

Hybrids:

The following hybrids have very silvery leaves:

'Berries and Cream'— pink, 12 inches.

'British Sterling'— pink to blue, 12 inches.

'Excalibur'— blue, 12 inches.

'Spilled Milk'— pink, 9 inches.

The following hybrids have nicely flecked foliage:

'Little White'— cobalt blue, 8 inches.

'White Wings' —white, 12 inches, mildew resistant to replace 'Sissinghurst White.'

Additional Notes:
Pulmonaria may win for the most common names. The choice includes: Bethlehem sage, lungwort, cowslip lungwort, Jerusalem sage, Joseph and Mary, spotted dog, soldiers and sailors.

Vinca minor

Common Periwinkle, Creeping Myrtle, Vinca
Apocynaceae (Dogbane Family)
○, ◑, ●   Ground cover, Early spring

Flowers: The flowers of common periwinkle add a wonderful touch to this foliage plant. The five-petalled, 1-inch solitary flowers are often violet-blue but can also be white, blue, or red-violet.

Leaves: Common periwinkle is best known for its leaves. The 1-inch leaves are opposite, glossy, dark green, and ovate with an obvious midvein. Some cultivars have variegated foliage.

Growing Needs: This plant will grow and thrive in a partially shaded, well-drained site. It can also grow in the full sun (with more flowers, but the leaves may yellow) as well as in heavy shade (with less flowers and dark glossy leaves).

Landscape Use:

Height:	4 to 6 inches
Habit:	Low mounded, spreading
Spacing:	12 to 18 inches (if weeded while filling in)
Texture:	Medium to medium fine

Periwinkle is an excellent ground cover and good for interplanting with bulbs. It also combines well in the front of the perennial border where there is plenty of space. Vinca is not a shy plant and will fill in and cover large spaces as well as non-vigorous plants.

Design Ideas: Common periwinkle makes an excellent interplanting foundation for growing the small bulbs, such as glory of the snow, snowdrop, and winter aconite, within the ground cover. Plant around the edge of shallowly rooted trees, such as star magnolia, and allow periwinkle to fill in underneath the tree.

Cultivars:

'Argenteovariegata' has leaves variegated with cream white and is less vigorous.

'Aureovariegata' has yellow variegation on the foliage.

'Atropurpurea' has purplish-red flowers.

'Bowles's Variety' produces larger violet blue flowers and is more clump forming.

'Gertrude Jekyll' has creamy white flowers.

'Multiplex' is a double-flowered purple.

'Plena' is a double-flowered blue.

Problems: None serious.

Propagation: Cuttings and division in spring.

Zones: 4-8

EARLY SPRING BULBS

Special Notes about Hardy Bulbs:

1) All are planted in fall, usually as soon as possible, in September or October for best results (root formation and flower bud development). A complete fertilizer (10-10-10) may be used.
The fall flowering bulbs *(Lycoris, Crocus, Colchicum)* are planted in summer or very early fall.

2) Plant the bulbs in groups of at least 10 to 12 (50 to 60 is great, too) for pleasing visual impact in the landscape.

3) Choose a well-drained, sunny or partially shaded site. A site under deciduous trees (ones that lose their leaves) works well because the early bulbs flower before the leaves appear. Avoid, however, sites with numerous or very shallow roots.

4) Do not plant in the front of the garden but, rather, place the bulbs in the foreground or even farther back in the garden. Perennials in the foreground or midborder will cover the ripening or yellowing (and browning) bulb foliage in the early summer garden. Good "cover" examples include hardy geraniums, daylilies, herbs, and ornamental grasses.

5) Allow the foliage to naturally ripen or yellow. Do not pull it out prematurely. The leaves continue to photosynthesize (manufacture food) and store food reserves in the bulb. Proper landscape placement in the middle of the garden, not in the front, will help you to avoid the temptation to pull out the fading and unsightly leaves. Just allow other taller perennials to naturally cover up the bulb foliage. It's less work!

6) Remove the faded flowers on tulips and hyacinths so seeds will not form. All other bulbs may be allowed to go to seed with no adverse effects on the bulbs. Most hybrid narcissus do not form seeds; the flowers will dry and eventually fall off.

Anemone blanda

Grecian Windflower, Greek Anemone
Ranunculaceae (Buttercup Family)
○, ◑   Foreground, Early spring

Flowers: The first daisy of the year! These showy daisy-looking flowers have 10 to 14 petaloid sepals that range in color from lavender-blue to pinks to white. The stamens are a distinct yellow and each solitary flower head is 1½ to 2 inches in diameter.

Leaves: The attractive, 3-inch, basal leaves are dark green, petioled, and divided into three segments. The leaves die back in early to midsummer.

Growing Needs: Grecian windflower prefers a full sun to partly shaded site. It does best in a well-drained soil and would benefit from a light mulch. Soak the tubers for several hours to overnight before planting.

Landscape Use:

Height:	3 to 6 inches
Planting depth:	3 inches
Spacing:	3 inches
Planting time:	Fall
Texture:	Medium fine

Mass Grecian windflower in drifts under deciduous trees. Or, plant windflower as a foreground planting behind a low edging.

Design Ideas: Plant the white flowering varieties with early red tulips. The bluish-lavender ones look great next to the blue-green foliage of pinks with the gray foliage of woolly thyme in front of both of them. Do not plant spring bulbs on the front edge of a garden due to the fading and browning foliage in summer; edging plants, such as thyme or pinks, will mask the yellowing foliage.

Cultivars:

'Blue Star' has blue flowers.
'Bridesmaid' has large white flowers.
'Bright Star' has bright pink flowers.
'Pink Star' has pink flowers.
'Radar' has magenta flowers with a white center.
'White Splendor' has white flowers.

Problems: None serious.

Propagation: Division after the plants (tubers) are dormant and seed (self sow to small degree).

Zones: 5-9

Chionodoxa luciliae

Glory of the Snow
Liliaceae (Lily Family)
○, ◑ Foreground, Early spring

Flowers: Each star-shaped, 1-inch flower has six violet to blue petals with a white center. Five to eight vivid flowers arise on a reddish stem. The flowering time can last up to two weeks.

Leaves: The 4- to 6-inch leaves of glory of the snow are linear and basal. They often appear only in twos and sometimes appear cupped or keeled at the tip. The foliage disappears by summer.

Growing Needs: Plant glory of the snow in the fall in full sun or partial shade in any well-drained site. This plant will self sow. The seeds can be collected and planted elsewhere in the garden. Leaves should be left until they start to turn yellow so the bulbs will replenish themselves.

Landscape Use:

Height:	4 to 6 inches
Planting depth:	3 inches
Spacing:	3 inches
Planting time:	Fall
Texture:	Medium fine

Glory of the snow looks excellent when planted en masse. It can be interplanted into shorter ground covers and performs well under deciduous trees.

Design Ideas: Glory of the snow is beautiful springing forth from the dark green leaves of English ivy. It is nice combined with early daffodils (Jack Snipe or February Gold) or Siberian squill or within low ground covers, such as ajuga or vinca. European wild ginger makes a great edging plant to show off glory of the snow.

Cultivars:

'Alba' has white flowers.
'Pink Giant' has pink flowers.

Problems: None serious.

Propagation: Division when plants go dormant or in fall (mark them for ease in finding the spot to dig) and seed by self sowing.

Zones: 3-9

Crocus chyrsanthus

Snow Crocus, Golden Crocus, Botanical Crocus
Iridaceae (Iris Family)
○, ◑ Foreground, Early spring

Flowers: Snow crocus flower earlier and have smaller flowers (1 inch) than the common crocus. Initially, and at night, each flower appears cup-like. After opening and during the day, each flower appears star-like. Each flower consists of six perianth segments and range in color from yellow, white, lavender, and bi-colors. Each flower has distinct yellow-orange stamens and some have brownish or purplish markings on the three outer perianth segments.

Leaves: The linear crocus foliage appears just at the end of flowering and will elongate to 8 to 10 inches. The obvious white midvein on the dark green linear leaf is a great identification feature until the foliage yellows and dies back in summer.

Growing Needs: These corms should be planted in fall. They will grow in any sunny well-drained site or in partial shade under deciduous trees. Let the leaves "ripen" after flowering to allow for the corm to replenish its food reserves for the next season. Leaves can be cut back after they have turned yellow.

Landscape Use:

Height:	4 to 6 inches
Planting depth:	3 inches
Spacing:	3 inches
Planting time:	Early fall
Texture:	Medium fine

Snow crocus makes an eye-catching display when planted in a mass of one color as a foreground plant, in rock gardens, or naturalized under trees or near ponds.

Design Ideas: Plant snow crocus behind low-growing lemon or silver thyme and sea pink and near early daffodils and *Geranium sanguineum striatum*, a lovely hardy geranium. Plant the early flowering crocus where they will be easily seen from the house or sidewalk during the late winter or early spring. Plant snow crocus within the lawn for some spring color. Do not mow the lawn until the foliage is yellowed, otherwise, crocus will only last a year or two rather than 10 or 12 years or more.

Cultivars:

'Prince Klaus' has deep purple flowers edged in white outside and completely white on the inside.

'Princess Beatrix' is sky blue and golden yellow at the base.

'Ladykiller' has deep purple on the three outer segments and opens white.

'Zwanenburg Bronze' has brown markings on the outer segments that contrasts with its golden color.

'Fuscotinctus' is a yellow crocus with violet stripes on the three outer segments.

Problems: None serious.

Propagation: Self sowing and by division when the corms are dormant.

The first spring after I moved to a new house, I returned to my little house and moved white *Crocus vernus* in full flower, removing a generous soil ball to keep the roots undisturbed. The replanted, well-watered crocus did fine and still flower every spring.

Zones: 3-8

Related Species:

Crocus vernus, Common Crocus, flowers later than the Snow Crocus. Its flowers are larger without the bi-color forms. A unique striped form ('Pickwick' or 'Striped Beauty') is available. They will also force well (flower early in containers). Many cultivars are available as follows:

'Flower Record' — purple

'Peter Pan' — white

'Yellow Mammoth' — yellow

C. sieberi, Sieber crocus, also flowers early along with snow crocus. Its purple to white petal-like segments have an orangish-yellow base or throat.

Eranthis hyemalis

Winter Aconite
Ranunculaceae (Buttercup Family)
○, ◑ Edging to foreground, Early spring

Flowers: Winter aconite has bright yellow buttercup-like flowers in early spring. Each cheerful solitary 1-inch flower has six petaloid sepals.

Leaves: Each flower has its own cheery 3-inch green collar of leaves. These leaves are palmately dissected and appear nearly circular beneath the flower. These glossy green leaves eventually die back in summer.

Growing Needs: A full sun to partially shaded, moist site is acceptable for growing winter aconite. For best results, plant in a well-drained, wind protected, cool shady area. Soak tubers overnight before planting them in early fall. It is very low maintenance — just plant them and forget them!

Landscape Use:

Height:	3 to 6 inches
Planting depth:	2 to 3 inches
Spacing:	3 inches
Planting time:	Early fall
Texture:	Medium fine

These bright little flowers are great used in masses or as an edging foreground plant. An effective use of this bulb is to interplant it in an existing ground cover, such as common periwinkle or English ivy. It naturalizes gracefully.

Design Ideas: Winter aconite combines perfectly with snowdrops planted within a carpet of English ivy. Place winter aconite in front of evergreen perennials or shrubs with low-growing campanulas, lamiums, or thymes in front of it. Since winter aconite flowers early, place it in an area where it can be easily viewed from a window or sidewalk. A location in front of hosta is perfect because the large hosta leaves will nicely cover the fading foliage in the summer.

Problems: None serious.

Propagation: Division and self sowing.

Zones: 3-7

Galanthus nivalis

Snowdrops
Amaryllidaceae (Amaryllis Family)
◯, ◑, ● Edging to foreground, Early spring

Flowers: The flowers of snowdrops are delicate, nodding, and white. They are 1-inch long and solitary (rarely in twos). The flower is made up of three outer oval segments and three fused inner segments that have green markings on the lower edges.

Leaves: Snowdrops produce two to three linear, basal, 4- to 8-inch leaves. The leaves grow with the flowers and continue to grow after flowering. The curved, leaf-like spathe above the pendulous flower is a good ID feature. The leaves are green to blue-green and eventually die down in early summer.

Growing Needs: These easy-to-grow plants prefer partial to full shade as well as full sun under deciduous trees. They do best in a cool, moist, well-drained location. You can plant them in the fall and forget them!

Landscape Use:

Height:	4 to 6 inches
Planting depth:	3 inches
Spacing:	3 inches
Planting time:	Early fall
Texture:	Fine

Snowdrops make an excellent foreground or "front of the border" plant when planted within evergreen ground covers. Snowdrops are the most effective when they are massed. Plant them to naturalize under trees.

Design Ideas: Snowdrops look nice just behind early spring edging plants, such as purple rockcress, rockcress, or basket of gold. While not in flower at the same time, the foliage of the edging plants will mask snowdrops' fading, yellowing leaves in early summer. They also look especially nice when planted with lavender snow crocus.

Cultivars:

'Flore Pleno' is a double-flower form.
'S. Arnott' ('Sam Arnott') is 10 inches and the outer segments are rounded rather than oval.
'Viridiapicis' has green tips on both the outer and inner floral segments.

Problems: None serious.

Propagation: Division in fall and self sowing.

Zones: 3-9

Related Species:

G. elwesii, giant snowdrop, has larger flowers and leaves and is 6 to 9 inches.

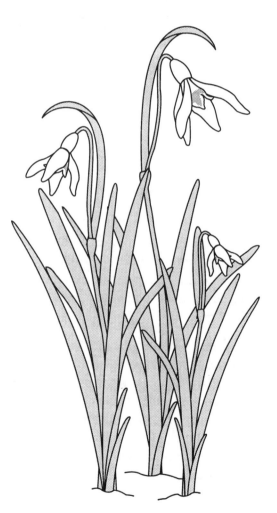

Hyacinthus orientalis

Dutch Hyacinth
Liliaceae (Lily Family)
◯, ◑   Foreground, Early spring

Flowers: Each flower stem is loaded with sweetly fragrant, flared, bell-shaped, ¼- to ½-inch flowers. These dense oval racemes range in color from white, blue, purple, pink, and salmon to yellow and may be 4 to 8 inches long.

Leaves: The 8- to 12-inch basal leaves are thick, 1 to 2 inches wide, and medium green. The tip of each parallel-veined leaf is cupped.

Growing Needs: Plant the bulb in the fall in a sunny, well-drained location. Plantings under deciduous trees are also effective. The bulbs (and flower size) decrease in size after the first year. The flower racemes become smaller and less full, which gives a less formal or more cottage-garden look. For formal sites, hyacinths may be treated as an annual.

Landscape Use:

Height:	8 to 12 inches
Planting depth:	6 inches
Spacing:	6 inches
Planting time:	Fall
Texture:	Bold

Dutch hyacinth looks best when massed in one color in the foreground. It also makes an excellent accent plant and can be forced for indoor color.

Design Ideas: Plant this bulb with daffodils in the foreground and with evergreen candytuft as an edging. Place pink flowering hyacinths in combination with lavender moss phlox and blue ajuga with bluish columbine foliage. White hyacinths and Virginia bluebells (under deciduous trees) make a nice duet.

Cultivars: Numerous!

'Amethyst' is a lilac color.

'Carnegie' is a pure white.

'City of Haarlem' is a late flowering yellow hyacinth.

'Delft Blue' is a beautiful porcelain blue flower.

'Jan Bos' is a cerise color.

'Lady Derby' is a fragrant pink.

'Pink Pearl' is a deep pink with paler edges on the florets.

Problems: None serious.

Propagation: Division.

Zones: 4-8

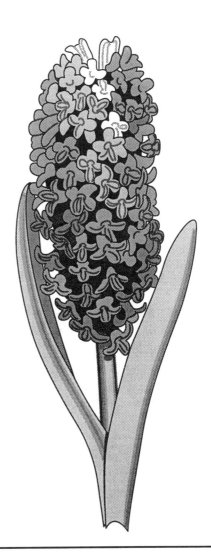

Iris reticulata

Netted Iris, Reticulated Iris
Iridaceae (Iris Family)
○ Edging, Early spring

Flowers: The blue to violet flowers of netted iris appear before its emerging foliage. Each solitary flower is composed of six segments, three upright standards and three downward falls with a conspicuous blotch of color (yellow, orange, or white) on each fall. The 2- to 3-inch flowers are sweetly fragrant and force well.

Leaves: The basal leaves of netted iris are very linear, angled, and green. After flowering, the leaves continue to elongate to 12 to 18 inches long and will then die back in summer.

Growing Needs: A sunny site is best for netted iris. The soil should be well drained.

Landscape Use:

Height:	4 to 6 inches
Planting depth:	3 inches
Spacing:	3 inches
Planting time:	Early fall
Texture:	Fine

Netted iris makes a great show as a small mass planting. It interplants well in shorter ground covers. Use it as an accent in the rock garden. This iris forces easily and is good for indoor color.

Design Ideas: Plant netted iris within common periwinkle beds or along with pasque flower, Siberian squill, and snow crocus. Combine it with the blue leaves of the shorter Dianthus or with the whitish snow in summer (*Cerastium*).

Cultivars:

'Cantab' is light blue with yellow and white accents.
'Harmony' is a deep blue with orange and white markings on the falls.
'Joyce' is a beautiful sky blue with orange markings.

'Natascha' is white with gold accents.
'Violet Beauty' is deep purple with orange crest.

Problems: "Ink disease" fungus causes mottling of leaves and eventually destroys the bulb.

Propagation: Division.

Zones: 4-8

Related Species:

Iris danfordiae, the Danford Iris, produces a canary yellow flower. This 4-inch iris needs a protected site and slightly acidic soil for a longer bulb life. It makes an excellent accent planting of 10 to 12 bulbs near rhododendrons.

Iris cristata, Crested Iris, is a 6-inch, early spring rhizomatous iris with fuller lavender blue flowers and flat, wider green foliage.

Leucojum aestivum

Snowflake
Amaryllidaceae (Amaryllis Family)
○, ◑ Foreground, Early spring

Flowers: The ½-inch, graceful, white snow-flake flowers are pendulous bells. Each petal is tipped with a green dot or marking. The flowers are borne in umbels of two to eight flowers.

Leaves: The linear leaves are 9 to 12 inches long and strap-like with rounded tips. The leaves resemble daffodil foliage but are green, not blue-green like daffodils. Snowflake foliage ripens (yellows) and dies back in summer.

Growing Needs: Any full sun to partially shaded location is suitable for snowflake. The soil should be well drained.

Landscape Use:

Height:	10 to 12 inches
Planting depth:	3 inches
Spacing:	4 inches
Planting time:	Fall
Texture:	Medium

Snowflakes make a good show in the foreground or the midborder. It is also appealing when naturalized near a woodland edge or pond.

Design Ideas: Snowflakes look lovely in almost any combination. Group it with basket of gold and late season red tulips or with daffodils and lady's mantle. For a partially shaded spot, place snowflakes with both fringed bleeding heart and old-fashioned bleeding heart with pansies as an edging.

Cultivars:

'Gravetye' or 'Gravetye Giant' is 18 to 20 inches tall and generally has larger and more numerous flowers per stem.

Problems: None serious.

Propagation: Division.

Zones: 4-9

Related Species:

L. vernum, spring snowflake, is an earlier flowering spring snowflake, 6 to 9 inches.

L. autumnale, autumn snowflake, flowers in the fall and is 9 to 12 inches tall.

Muscari armeniacum

Armenian Grape Hyacinth
Lilicaeae (Lily Family)
○, ◑   Foreground, Early spring

Flowers: The grape-like clusters (racemes) of blue-violet flowers are pendulous and fragrant. Each ¼-inch flower is rounded or grape-like at first and then opens to a bell shape, often with a white tip. Seed heads may be trimmed back if self sowing is not desired.

Leaves: The evergreen, narrow, basal leaves emerge in late summer or early fall after planting. The medium green leaves will overwinter (with some browning of the leaf tips) and then die back in early summer after flowering.

Growing Needs: This plant thrives in a full sun to partially shaded location. Grape hyacinth needs a well-drained soil. It will self sow and colonize an area easily.

Landscape Use:

Height:	6 to 9 inches
Planting depth:	3 inches
Spacing:	3 inches
Texture:	Medium

Armenian grape hyacinth looks best when massed in the foreground of the border. It also makes an excellent companion plant in ground covers and looks good when naturalized under deciduous trees.

Design Ideas: Try interplanting this plant with pink moss phlox, bugleweed (*Ajuga*), perennial flax (*Linum*), or sweet woodruff (*Galium*). Grape hyacinths combine well with tufted sea pink (*Armeria*) as an edging and catmint (*Nepeta*) as a midborder plant behind it.

Cultivars:

'Blue Spike' has double flowers.

'Dark Eyes' has bright blue flowers, rimmed with white.

'Heavenly Blue' has vivid blue flowers.

Problems: None serious.

Propagation: Self sows and division.

Zones: 3-8

Related Species:

Muscari botryoides, common grape hyacinth, is grown for its white variety — var. *album*, which is less vigorous than the blue-violet ones but very attractive!

M. comosum '*Plumosum*' produces feathery fringed plumes of purple flowers.

M. latifolium has a two-toned flower, the upper part is the regular blue-violet color and the lower part is a dark blue.

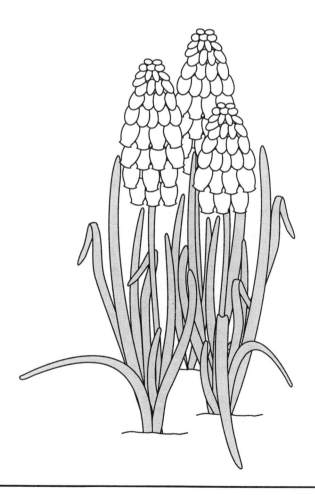

Narcissus

Daffodil, Narcissus
Amaryllidaceae (Amaryllis Family)
○, ◑   Foreground, Midborder, Early spring

Flowers: The flowers of daffodils vary greatly. Each flower has a central cup or trumpet-shaped corona surrounded by a perianth. The corona may be at 90-degree angles to the perianth or flattened against it. Narcissus may be borne solitary or in multiples. The flowers come in various colors — white, yellows, oranges, bicolors, and sometimes, pink or green. Narcissus have been classified into 12 divisions. Plant a variety of types to extend the flowering time of daffodils as follows:

Earliest (March–early April)

Cyclamineus (Div. 6)
Species (Div. 10)

Early (early April)

Trumpets (Div. 1)
Long cup (Div. 2)

Mid (mid April)

Small cup (Div. 3)
Double (Div. 4)
Triandrus (Div. 5)
Split corona (Div. 11)

Late (late April)

Jonquilla (Div. 7)
Tazetta (Div. 8)
Poeticus (Div. 9)

Leaves: The basal foliage is linear, rounded at the tip, and bluish green. Some divisions have narrow reed-like foliage.

Growing Needs: Daffodils do well in a sunny, well-drained location. A site under deciduous trees is also effective because it is in full sun when the daffodils are in flower. After flowering, allow the foliage to yellow and then cut it back. Removing faded flowers is optional because most types will not develop seeds. Remove if any seeds are being formed.

Landscape Use:

Height:	3 inches to 2 feet
Planting depths:	Larger ones—6 inches (8 inches in sandy soils) Miniatures—3 inches
Spacing:	6 inches 8 inches—Naturalizing Miniatures—3 inches
Planting time:	Fall

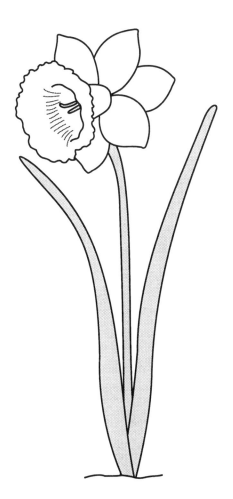

Texture:	Medium fine to medium bold

Plant daffodils in groupings in the foreground to midborder. A cluster of (at least) 12 to 15 is recommended for visual impact. Plant hundreds under trees along a lane and allow them to naturalize. (Delay mowing until the leaves yellow.) Daffodils give the best value of all the bulbs for the money spent because they are easy to grow, increase each year, and last for years and years.

Design Ideas: Interplant the earliest daffodils within ground covers, such as common periwinkle or ivy. Try combining daffodils with pansies, rockcress, and evergreen candytuft. A favorite inter-planting combination is daffodils within a well-spaced (18 inches apart) daylily bed.

Divisions and Characteristics:

Div. 1 — Trumpet

The trumpet is as long as, or longer than, the perianth, having one flower per stem. 'Dutch Master', 'Mount Hood', 'Golden Harvest'

Div. 2 — Long Cup

The cup length is more than $1/3$, but less than equal to the length of the perianth segments, having one flower per stem. 'Carlton', 'Flower Record', 'Ice Follies', and "pinks" such as 'Mrs. R.O. Backhouse', 'Salome'

Div. 3 — Small Cup

The cup length is not more than $1/3$ the length of the perianth segments, having one flower per stem. 'Edward Buxton', 'Polar Ice'

Div. 4 — Double

Solitary flowers with double flower parts, either the cup or the perianth. 'Texas', 'Golden Ducat'

Div. 5 — Triandrus

Clusters of fragrant pendulous bell-shaped flowers often with reflexed (flung back) perianth segments. 'Thalia', 'Ice Wings', 'Liberty Bells'

Div. 6 — Cyclamineus

Early miniature trumpet or long-cupped flowers with reflexed perianth segments.

'February Gold', 'Jack Snipe', 'Peeping Tom', 'Tête a Tête'

Div. 7 — Jonquilla

Clusters of fragrant flowers, having a perianth at a 90-degree angle (flat). Foliage is slender and reed-like. 'Suzy', 'Pipit', 'Trevithian'

Div. 8 — Tazetta

Clusters of fragrant shallow-cupped flowers.

'Cheerfulness', 'Yellow Cheerfulness', 'Geranium' Paperwhites are not hardy but excellent for forcing (and overly fragrant).

Div. 9 — Poeticus

Single-flowered fragrant plants with tiny cups in a range of colors and a white perianth. 'Actaea', 'Old Pheasant's Eye'

Div. 10 — Species

Wild forms and their variations.

Bulbocodium conspicuus — 'Yellow Hoop Petticoat'

Div. 11 — Split Corona (Butterfly Narcissus)

Corona (usually bicolor) is split into six parts and flares out against the perianth. 'Baccarat', 'Marie-Jose', 'White Butterfly'

Div. 12 — Miscellaneous

Varieties of garden origin that do not fit any other division.

Problems: None serious.

Propagation: Division.

Zones: 4-8

Pulsatilla vulgaris

(Anemone pulsatilla)

Pasque Flower
Ranunculaceae (Buttercup Family)
○, ◑ Foreground, Early spring

Flowers: The solitary 1½- to 2½-inch flowers are purple, red, and creamy white with showy yellow stamens in the center. The outer petals of pasque flower are covered with long white pubescence — "a must to pet!" The seed head looks like a shaggy fuzzy ball.

Leaves: The 4- to 6-inch long, finely dissected leaves, as well as the stems, are also extremely pubescent. The ferny foliage emerges after the flowers and is attractive all season long. This plant is delightfully soft to the touch.

Growing Needs: Pasque flower grows well in partial shade to full sun in any well-drained soil. The tubers should be soaked before planting.

Landscape Use:

Height:	12 inches
Planting depth:	2 inches (usually purchased as a plant)
Spacing:	12 inches
Planting time:	Spring or fall
Texture:	Medium to medium fine

Plant a single pasque flower in the foreground of the flower border as an excellent accent. Groups of two or three are also pleasing. It also makes a great addition to the rock garden.

Design Ideas: Combine pasque flower with daffodils and evergreen candytuft. For partial shade, place it with ferns, bleeding hearts, and forget-me-nots. It is an excellent accent in sweet woodruff.

Cultivars:

'Alba' has a 6- to 10-inch white flowered plant.

'Bartons Pink' has a pink flower and is 6 to 10 inches tall.

'Rubra' is a dark red, 6- to 10-inch plant.

Problems: None serious.

Propagation: Division, seed, and root cuttings.

Zones: 5-8

Puschkinia scilloides

Striped Squill
Liliaceae (Lily Family)
○, ◐   Foreground, Early spring

Flowers: Striped squill produces five to twenty florets per raceme. Each ½-inch, star-shaped to bell-shaped flower is whitish blue with a deeper blue stripe down the center of each petal. The lightly fragrant flowers may face outward to upward.

Leaves: The basal foliage is 4 to 6 inches, linear, and medium green. The leaves will die back by summer.

Growing Needs: This plant performs well in full sun to partial shade in any well-drained soil. Plant it; forget it.

Landscape Use:

Height:	6 inches
Planting depth:	3 inches
Spacing:	3 inches
Planting time:	Early fall
Texture:	Medium fine

Use striped squill in masses in the foreground of the flower bed. It also looks equally well in rock gardens or naturalized along woodland edges or in an informal lawn setting.

Design Ideas: Group this plant with low-growing sedums and pinks. Striped squill makes a nice spring display with Siberian squill and early daffodils. Plant woolly thyme or snow in summer in front of these bulb groupings.

Cultivars:

'Alba' has pure white flowers.

var. *libanotica* has smaller flowers with more sharply acute petals.

Problems: None serious.

Propagation: Division and self sowing.

Zones: 3-10

Scilla sibirica

Siberian Squill
Liliaceae (Lily Family)
○, ◑, ● Edging, Early spring

Flowers: The flowers of Siberian squill are ½ inch in size and have pendulous flowers with an occasional upturned star-shaped one. They are blue with a darker stripe on the outside of the petals. Each stem has one to six flowers per raceme.

Leaves: The linear leaves of Siberian squill are 4 to 6 inches long, basal, and green. The foliage will wither away by summer.

Growing Needs: Siberian squill can easily be grown in full sun or in full shade. It performs best in a well-drained soil and is one of the easiest bulbs to grow. I have witnessed them happily growing in a crack in the sidewalk.

Landscape Use:

Height:	4 to 6 inches
Planting depth:	3 inches
Spacing:	3 inches
Planting time:	Early fall
Texture:	Medium fine

Siberian squill looks best when naturalized or planted in groupings. It is also nice when planted just behind the edging plants in the perennial border.

Design Ideas: Plant Siberian squill with glory of the snow and early daffodils for a cheery spring grouping. Or, combine it with striped squill and the soft ferny foliage of pasque flower. I have planted it in masses in the center of an island bed with hardy geraniums, obedient plant, and campanulas to cover its fading foliage.

Cultivars:

'Alba' has a white flower.
'Spring Beauty' is a larger flowering form with long-lasting, scented flowers.

Problems: None serious.

Propagation: Self sowing and division.

Zones: 3-9

Tulipa

Tulip
Liliaceae (Lily Family)
○, ◑　Foreground, midborder
Early spring to late spring

Flowers: The cup-shaped flowers of tulips come in just about every color except a true blue. Besides the many colors, there are many petal characteristics — pointed, fringed, twisted, striped, and feathered. The flowers are either single, semi-double or double and are solitary but also occasionally have multiple flowers per stem.. Since there are so many distinctive shapes and characteristics, the tulips have been classified into types. For landscape planning, it is easiest to learn them by flowering time as follows:

Very Early (late March–early April)

T. kaufmanniana (Waterlily Tulip) 4 to 12 inches, 'Shakespeare', 'Stresa', 'Waterlily'

T. greigii — mottled foliage, 6 to 20 inches, 'Red Riding Hood', 'Plaisir'

T. fosteriana (Emperor types) 12 to 15 inches, 'Red Emperor', 'White Empress'

Early (mid April–early May)

Single early — fragrant, 10 to 18 inches, 'Princess Irene', 'Couleur Cardinal'

Double early — resembles a peony, 10 to 12 inches, 'Dante', 'Viking', 'Monte Carlo'

Single Early

Double Early

Emperor Type

Lily-flowered

Darwin

Parrot

T. praestans — multiple flowers/stem 8 to 12 inches, 'Fusilier'

Mid Season (late April–mid May)

Mendel — forces well, 14 to 24 inches, 'Apricot Beauty', 'High Society'

Triumph — sturdy stems, bicolors, 18 to 24 inches, 'Arabian Mystery', 'Dreaming Maid', 'Merry Widow', 'Shirley'

Darwin Hybrids — large flowers, sturdy, 2 feet, 'Beauty of Apeldoorn', 'Dover', 'Olympic Flame', 'Pink Impression'

T. tarda — star-like, yellow with white edge, 4 to 6 inches, strappy foliage

Late Season

Darwin — well-loved, large flowers, 24 to 30 inches, 'Queen of the Night' (almost black), 'Dreamland', 'Renown', 'Twinkle'

Cottage — lovely egg-shaped flowers, 16 to 30 inches, 'Golden Harvest', 'Greenland'

Double late — peony-flowered, 18 to 24 inches, 'Angelique', 'Lilac Perfection'

Rembrandt — streaked or flecked petals, 2 feet, 'Gloire de Holland'

Lily-flowered — striking pointed petals, flowers resemble lilies when open, 18 to 26 inches, 'Marilyn', 'Queen of Sheba', 'West Point'

Parrot — large fringed, ruffled petals in wild colors, 20 to 24 inches, 'Estella Rijnveld', 'Flaming Parrot.' The 'Black Parrot' was the first tulip that I grew when I was 12 — striking!

Leaves: The leaves of tulips are basal or alternate on the stem. Each wide, pointed leaf is medium-green and broad ovate. Some types are mottled with burgundy coloration

Growing Needs: Tulips prefer a well-drained, sunny location. A site under deciduous trees is also suitable. Plant the bulbs in fall. At planting time, add a balanced fertilizer to the soil for best results. In summer, allow foliage to yellow and then cut it back. Remove faded flowers to prevent seed formation.

Landscape Use:

Height:	4 to 28 inches
Planting Depth:	6 inches
	4 inches (Early ones)
Spacing:	4 to 6 inches
Planting time:	Fall
Texture:	Medium fine to medium bold

Tulips look great in the formal garden when planted en masse. A small grouping (12 to 15) makes an excellent accent in the foreground to midborder. Although tulips are not as long-lived as daffodils, they do add a needed showy lift to the spring garden.

Design Ideas:

Early ones — Interplant the early birds near the house with in ground covers. Red Emperor (*T. fosteriana)* combines well with evergreen candytuft and coral bells. Red Riding Hood tulips (*T. greigii*), woolly thyme and cushion spurge form a lovely trio in the early spring garden.

Later ones — Plant them around lush hardy perennials (fall anemones, artemisias, asters, Shasta daisies) that can mask the withering tulip foliage. Pandion (lavender) tulips look great with *Phlox divaricata* and epimediums. I really enjoy the shape of the lily-flowered tulips with the mounded lady's mantle in the front and the mounded baptisia behind them.

Cultivars: There are many spectacular cultivars available.

Problems: None serious.

Propagation: Division and seed.

Zones: 5-8 (In the south, bulbs must be given a cold treatment before planting. Tulips are treated as an annual in the south.)

SPRING PERENNIALS

Ajuga reptans

Bugleweed, Ajuga
Lamiaceae (Mint Family)
○, ◐, ●   Ground cover, Spring

Flowers: The flowers of bugleweed are borne in whorls on leafy spikes. Each showy blue to purple (also pink and white) flower is 1 inch in diameter and irregular in shape.

Leaves: The 2-inch basal leaves are oval or oblong and a dark glossy green or purple. The stem leaves are smaller and also interspersed within the flower spikes. The leaves are commonly evergreen.

Growing Needs: What an easy-to-grow plant for full shade to full sun! (Almost too easy.) It will flourish in any well-drained soil. Give this plant room to spread.

Landscape Use:

Height:	6 to 10 inches
Habit:	Creeping, low mounded
Spacing:	12 to 18 inches
Texture:	Medium

Bugleweed is best used as a ground cover. It is excellent planted in hard-to-grow areas, such as under trees. Use bugleweed as a confined edging plant; it will spread into the lawn area. Divide or transplant the clump to keep it the preferred size. Occasionally large areas of ajuga may appear spotty, with small missing sections of it due to crown rot.

Design Ideas: Plant bugleweed with pink creeping phlox. Interplanting with daffodils makes a nice contrasting combination. Columbines, soapwort (*Saponaria*), and miniature roses make good companions with ajuga. It also performs well under crabapples or serviceberries.

Cultivars and Related Species:

'Alba' has white flowers and light green leaves.

'Burgundy Glow' has interesting green and creamy white foliage with burgundy new growth, blue flowers.

'Catlin's Giant' has large bronze to green leaves, blue flowers, grows 8 to 10 inches.

'Jungle Beauty' has interesting large multi-colored leaves of purplish green, cream, and reddish pink. 'Jungle Beauty Improved' has large purple leaves and grows to 12 inches.

'Mini Crispa Red' is a low-growing (4 to 5 inches), crinkly, reddish-leaved ajuga with blue flowers.

A. genevensis 'Pink Beauty' has pink flowers, more clump forming.

A. pyramidalis 'Metallica Crispa' has dark purple, crinkled leaves, blue flowers.

Problems: Crown rot, divide it every three to five years and plant in well-ventilated areas. Treat with a fungicide when needed.

Propagation: Division in spring or fall and cuttings.

Zones: 4-8

Alchemilla mollis

Lady's Mantle
Rosaceae (Rose Family)
◑, ○   Edging, Foreground, Spring

Flowers: Lady's mantle produces a mass of ¼-inch, greenish yellow flowers in a compound cyme. The fine-textured flowers are very attractive and borne just above the bold foliage. It makes a great cut flower, fresh and dried.

Leaves: Each gracefully toothed basal leaf is 4 to 6 inches wide, rounded, and gently lobed. The surface is softly pubescent and light green. Each new leaf unfolds in a pleated fashion like a woman's mantle or coat. In the morning dew or after a rain, beads of water collect on the leaves and look extremely attractive. The leaves will remain attractive all through the growing season if sited well.

Growing Needs: These plants prefer moist partial shade, but will tolerate full sun if kept moist. The site should be well drained and fertile. Add organic matter and generously mulch to increase the moisture-holding capacity of the soil.

Landscape Use:

Height:	12 inches
Habit:	Mounded
Spacing:	18 inches
Texture:	Medium fine in flower
	Bold in leaf

Lady's mantle makes a lovely shady edging or foreground planting. It can be used as a limited ground cover. Lady's mantle also makes an excellent specimen planting. I have grown it for 10 years in full sun in a well-mulched bed and it looks better every year.

Design Ideas: Lady's mantle looks great with deep blues like perennial blue salvia, delphinium, or veronica. Spring bulbs, such as snowflake and lily-flowered tulips (white or reddish Queen of Sheba), look great behind it. I enjoy it planted with catmint because the fine-textured flowers blend gracefully together.

Problems: None serious.

Propagation: Division and by seed.

Zones: 4-7

Related Species:

A. alpina is a compact 4 to 6 inch plant with more distinct lobed leaves. Attractive!

Amsonia tabernaemontana

Blue Stars
Apocynaceae (Dogbane Family)
◯, ◑ Midborder, Spring

Flowers: These ½-inch, star-like flowers are light blue and borne in dense terminal clusters. The five-petalled flowers are very open and strappy.

Leaves: The leaves are 3 to 6 inches long, alternate, and glossy. Each leaf is lanceolate (lance-shaped) appearing willow-like in shape. In the fall, the leaves turn a lovely golden yellow.

Growing Needs: This low-maintenance plant will grow in either sun or partial shade. The habit is more sprawling in partial shade and it may need peastaking. Any well-drained soil of average fertility will suffice. Blue stars will also tolerate a drought.

Landscape Use:

Height:	24 to 36 inches
Habit:	Upright mounded or vase shaped
Spacing:	24 inches
Texture:	Medium fine

Blue stars look best when massed in small groups in the midborder near bolder-textured plants as a good contrast.

Design Ideas: Combine blue stars with vivid blue *Veronica* 'Crater Lake Blue' or perennial blue salvia and *Silene* 'Robin's White Breast.' Try the combination of pink astilbe, Lady lavender, and snow-in-summer with blue stars behind it. A mass planting of blue stars is effective for holding down erosion on slopes and will require little maintenance or division.

Variety: var. *salicifolia* has a more narrow leaf than the straight species.

Problems: None serious.

Propagation: Division, seed (will self sow), and cuttings.

Zones: 3-9

Aquilegia hybrids

Columbine
Ranunculaceae (Buttercup Family)
○, ◐ Midborder, Spring

Flowers: These lovely flowers are rather unique. Each 2- to 3-inch flower has five petals and five sepals. Each petal has a backward projecting spur, which gives this flower its unique look. There are many colors—even bi-colors and doubles in reds, pinks, whites, yellows, purples, and blues.

Leaves: The attractive leaves are in threes, that is, two to three ternately compound. The light green 1- to 2-inch leaflets are rounded, gently lobed, and delicately lacy looking. Most leaves are basal with the stem leaves arranged alternately.

Growing Needs: Columbine prefers partial shade but can withstand the full sun if kept mulched and moist. Well-drained soil is a must and should be moderately fertile. If this plant is not sited well, it will be short-lived.

Landscape Use:

Height:	18 inches to 3 feet
Habit:	Two tiered (Upright flowers over mounded foliage)
Spacing:	12 inches
Texture:	Medium fine

Columbine makes an excellent specimen plant. It is also effective as a massed foreground to midborder planting.

Design Ideas: In partial shade locations, plant columbine with ferns and foxgloves. In a sunny location, plant columbine with coral bells, perennial flax, and bugleweed.

Columbines look great interplanted with tulips and daffodils (they will mask the fading foliage) along with golden sage, pinks, and soapwort as the edging plantings. For a lovely old-fashioned look, plant columbines with Shasta daisies, garden iris, and Siberian iris in front of peonies and tall delphinium.

Cultivars:

'Biedermeier' is a short-spurred, compact (12 to 18 inch), sturdy plant with a large color range of solids or bicolors.

'Crimson Star' is a red and white bicolor, 2 to 2½ feet, and self sows true to type.

'McKana Giants' are showy, long-spurred, tall (2½ feet), and bicolored.

'Music' hybrids are compact, 1½ feet, with large long-spurred flowers, coming in a wide range of colors.

'Snow Queen' has a pure white flower.

Other Hybrid Mixes:

Blackmore and Landgon, Blue Shades, Dragonfly Hybrids, Dynasty Series, Harbutt's Hybrids, Harlequin Mixed, Langdon's Rainbow Hybrids, Lowdham Strain, Spring Song Mixed

Variegated Ones:

'Leprechaun Gold' —purple flowers, 2 to 2½ feet.

'Lime Frost' — mixed colors, 2 feet.

Aquilegia vulgaris 'Variegata' — light blue flowers, 18 inches.

Problems:
Columbine is susceptible to leaf miner. Remove any leaves with leaf miner immediately and destroy to prevent further spread. If leaf miner is wide spread, the plant can be cut back after flowering to force new uninfected growth. Other less frequent problems include aphids, columbine borer, leaf spots, and rots (crown or root).

Propagation:
Seed (collect it yourself), transplant self-sown seedlings, and early fall division.

Zones:
3-9

Related Species:

Aquilegia vulgaris (European Columbine or Granny's Bonnet) is short-spurred, 2 feet, and less susceptible to leaf miner.

A. vulgaris 'Nora Barlow' has double flowers with no spurs, red petals (and some green) tipped in white, 2½ to 3 feet.

A. chrysantha is long-spurred, pale yellow, and grows 2 feet. 'Maxi'—longer flowering.

'Nana' golden flowers and dwarf habit.

'Yellow Queen' has profuse bright yellow, long-spurred flowers, 3 feet.

A. flabellata (Fan Columbine) is compact (8 to 10 inches) with nodding flowers and lovely blue to blue-green foliage.

var. *pumila* 'Ministar' — blue and white.

'Alba' — white.

Armeria maritima

Sea Pink, Thrift
Plumbaginaceae (Plumbago Family)
○ Edging, Spring

Flowers: These 1-inch flower heads are pink or white and borne on a leafless stem above the foliage. The cymose flower head is globular or rounded and resembles a smaller chives in shape (but not color). Faithful pinching back of the faded flowers will reward you with sparse flowering throughout the summer.

Leaves: The green foliage grows from a rosette and is very thin and narrow. These tufted clumps of 4-inch narrow leaves resemble a lush turf.

Growing Needs: Sea pink is a hardy full-sun plant. It does well in any well-drained soil and tolerates dry, infertile ones quite well. When the center of a large clump dies out, divide the plant. Pinch back spent flowers for reflowering.

Landscape Use:

Height:	4 to 8 inches
Habit:	Mounded or two tiered
Spacing:	9 to 12 inches
Texture:	Fine

Sea pink looks great in the front of the border as a single specimen or in groups. It can be used as a limited ground cover or in the rock garden. This clump former should not be interplanted with bulbs because of its dense growth; place bulbs behind it.

Design Ideas: Sea pink combines well with variegated lemon thyme as well as catmint and penstemon as a backdrop. *Armeria* looks absolutely fabulous with *Iris pallida* 'Argentea Variegata'. It makes a cheerful grouping when planted with cushion spurge, daffodils, and white or pink moss phlox.

Cultivars:

'Alba' has a white flower and is 5 inches.

'Bee's Ruby' is bright pink, 18 inches.

'Corsica' has a salmon pink flower, 6 inches.

'Dusseldorf Pride' is deep pink, 8 inches.

'Laucheana' is 4 to 6 inches and a deep rosy red, reblooming if deadheaded.

'Splendens' is bright rosy pink, 8 inches.

Problems: None serious.

Propagation: Seed and division.

Zones: 4-8

Related Species:

A. juniperifolia has beautiful tufted blue-green foliage with rose pink flowers and grows a compact 2 to 4 inches.

A. arenaria (plataginea) is a taller (1 to 2 feet) thrift with wider foliage. 'Kieft's Pastels' are various pastels — lavender to pinks and grow 18 inches.

Asarum europaeum

European Wild Ginger
Aristolochiaceae (Birthwort Family)
◐, ●   Ground cover, Spring

Flowers: The 1-inch flowers of European wild ginger are inconspicuous because they are borne under the leaves near the soil surface. Each maroon-brown flower is bell-shaped and three-lobed.

Leaves: This plant is prized for its elegant glossy foliage. The 3-inch diameter leaves are kidney-shaped, long-petioled from the rhizome, and usually evergreen.

Growing Needs: Moist partial shade to deep shade is appropriate to site this plant. The soil should be well drained, slightly acidic, and high in organic matter. Mulching is beneficial.

Landscape Use:

Height:	6 inches
Habit:	Low spreading
Spacing:	8 inches
Texture:	Medium

The foliage of this plant makes it a lovely ground cover in a shady garden. It looks great under trees as the only planting.

Design Ideas: This excellent high-interest ground cover goes well with heartleaf bergenia, primroses, and miniature daffodils. European wild ginger also looks nice with Bethlehem sage and shooting star. Do not interplant it with bulbs due to the dense covering of the soil by the rhizomes.

Problems: None serious.

Propagation: Division in spring or fall

Zones: 4-7

Related Species:

Asarum canadense (Common Wild Ginger), a native species, produces a pubescent 6-inch diameter leaf that is deciduous. This plant grows well in slightly acid to slightly alkaline soils.

A. shuttleworthii (Mottled Wild Ginger) has deep green foliage mottled with silver and grows 3 to 4 inches tall.

Aster alpinus

Alpine Aster
Asteraceae (Sunflower Family)
○, ◑ Edging, Foreground, Spring

Flowers: This daisy-like flower head has ray flowers in pinks, white, lavender, and blue with a center of yellow disc flowers. Each solitary flower head is 1 inch in diameter.

Leaves: The 1- to 2-inch basal leaves are pale green, oblong, and evergreen; the stem leaves are alternate, linear and somewhat smaller.

Growing Needs: Any sunny location will suit alpine aster. Asters tolerate partially shaded conditions. The soil must be light, very well-drained, and slightly alkaline.

Landscape Use:

Height:	10 inches
Habit:	Mounded
Spacing:	10 to 12 inches
Texture:	Medium

Alpine aster is wonderful for the sunny to partially shaded foreground. It is excellent used as a facer plant to cover up "leggy" plants (some gaillardias or anchusas).

Design Ideas: Alpine aster looks nice when planted with bugleweed and columbines. Another combination is alpine aster with snow-in-summer and *Veronica* 'Sunny Border Blue'. Alpine aster with soapwort (*Saponaria*) make a nice duo whether in monochromatic pink or in blending blue (of the aster) and pink (of the soapwort).

Cultivars:

'Goliath' has a lavender-blue flower.

'Dark Beauty' has deep blue flowers.

'Happy End' is a semi-double lavender and grows to 1 foot.

'Roseus' has pink flowers and is 6 inches tall.

Problems: Aster yellows, mildew, and rust.

Propagation: Seed, division, and cuttings.

Zones: 4-9

Baptisia australis

Blue Indigo, False Indigo, Blue False Indigo
Fabaceae (Legume Family)
○ Background, Spring

Flowers: Each individual 1-inch flower is blue and pea-like, having wings and a keel. Many florels are borne in a terminal raceme, which can be 9 to 12 inches long. After the plant flowers, it produces interesting 2- to 3-inch, inflated-looking black seed pods that persist throughout the winter and are decorative in dried arrangements.

Leaves: The attractive blue-green foliage is alternate and has three leaflets with a pair of noticeable stipules. Each leaflet is ovate and short-pointed and 1 to 3 inches long. Blue indigo remains attractive all season until blackened by severe frosts. While not evergreen, it has a dramatic winter character.

Growing Needs: This drought-tolerant plant will thrive in full sun and will tolerate a lightly shaded site but be more sprawling in habit. Blue indigo is very low maintenance and requires no division; it increases and looks better every year. Cutting back the old stems in early spring is the only maintenance required. Small seedlings or young plants can be transplanted; otherwise, site *Baptisia* well and leave it there to flourish.

Landscape Use:

Height:	3 to 5 feet
Habit:	Vase shaped
	Mounded with age
Spacing:	2 to 3 feet
Texture:	Medium to medium bold

Blue indigo makes an excellent background planting in the perennial border or used as a specimen planting. It can also be used in the cut flower garden for its flowers and seed pods.

Design Ideas: Plant blue indigo with peonies, oriental poppies, and irises. Try grouping it with li-lac cranesbill and snowflakes (*Leucojum*). Plant it in the center of an island bed with ornamental grasses, such as feather reed grass *(Calamagrostis),* and the tall *Rudbeckia nitida* 'Herbstonne' (black-eyed Susan 5 to 6 feet).

Problems: None serious.

Propagation: Tip cuttings, transplant small seedlings, and by seed.

Zones: 3-9

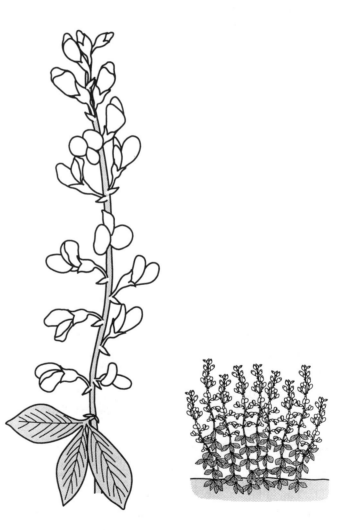

Brunnera macrophylla

(Anchusa myosotidiflora)

Perennial Forget-Me-Not, Siberian
   Forget-Me-Not, Siberian Bugloss
Boraginaceae (Borage Family)
◑ Edging, Foreground, Spring

Flowers: These dainty intensely blue ¼-inch flowers are five-petalled, daisy-like, resembling the classic forget-me-not (*Myosotis sylvatica*), and are borne in loosely branched racemes.

Leaves: The 6- to 8-inch basal leaves are heart-shaped, long petioled, roughly hairy, and dark green. The stem leaves are alternate, smaller, sparser, and either shorter petioled or sessile (no petiole).

An interesting note is the leaves enlarge after the flowering period.

Growing Needs: Partial shade will support the best growth. Perennial forget-me-not tolerates dry shade. In sunnier sites, the soil should be moist, yet well drained, and high in organic matter. As with most perennials, divide it when the center of the clump dies out.

Landscape Use:

Height:	12 to 18 inches
Habit:	Mounded
Spacing:	18 inches
Texture:	Bold to medium bold

Use perennial forget-me-not as a specimen planting or in small groupings near the garden edge or in the foreground position. It also makes a nice limited groundcover.

Design Ideas: Perennial forget-me-nots look especially nice with yellows, such as barren strawberry (*Waldsteinia*) in front or leopard's-bane (*Doronicum*) behind it. *Brunnera* combines well with European wild ginger, daffodils, bleeding hearts, and globe flower (*Trollius*) and lends a nice bold leaf texture to the garden.

Cultivars:

'Variegata' or 'Dawson's White' is variegated with a creamy white edge.

'Hadspen Cream' also has a creamy variegation but is more tolerant of uneven moisture.

Problems: Slugs.

Propagation: Division in spring, root cuttings, and seed (limited self sowing).

Zones: 3-8

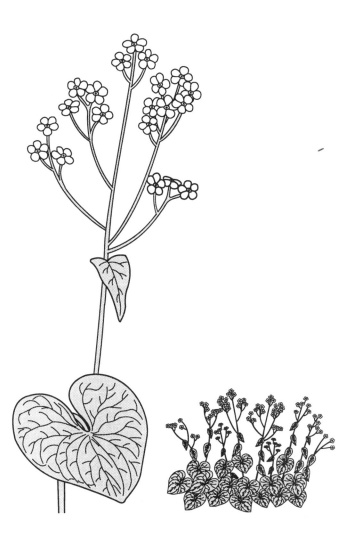

Centaurea montana

Mountain Bluet, Perennial Bachelor Button or
    Cornflower, Knapweed
Asteraceae (Sunflower Family)
○, ◑ Midborder, Spring

Flowers: These striking, solitary blue flower heads are 2 to 3 inches in size. The ray flowers are very thin and flared at the tips, making the flower appear lacy and very open. Each flower head is subtended by papery bracts. To prevent self sowing, deadhead after flowering.

Leaves: The lanceolate leaves of mountain bluet are 5 to 7 inches long and covered with stiff bristly pubescence. The plant has both basal leaves with petioles and stem leaves, which are alternate, sessile, and decrease in size near the flower. Newly emerging leaves appear whitish or silvery.

Growing Needs: This drought-resistant plant is best sited in a full-sun location but will tolerate light shade (growth will be more straggly). If cut back after flowering, there may be some sparse reflowering. Taller plants may need staking. Frequent division will produce a plant with a more pleasing habit.

Landscape Use:

Height:	1 to 2 feet
Habit:	Sprawling mounded
Spacing:	12 inches
Texture:	Medium
	Flowers— medium fine

Mountain bluet works well massed in the sunny midborder. Because it can sometimes look untidy, plant it with mounded facer plants.

Design Ideas: Mountain bluet looks excellent with yellow flowering plants, such as sedums (*S. kamtschaticum*) or columbines. It also blends well with snow-in-summer. Good facer plants are alpine aster and hardy geraniums (*G. sanguineum*).

Cultivars:

'Alba' has white flowers.
'Carnea' has pink flowers.
'Rubra' has dark rose flowers.
'Violetta' has dark purple flowers.

Problems: None serious.

Propagation: Seed (collect your own seed before it self sows) and division.

Zones: 3-9

Related Species:

C. hypoleuca 'John Coutts' — fuller lavender flower with lovely lobed leaves, 2 feet.

Cerastium tomentosum

Snow-in-Summer
Caryophyllaceae (Pink or Carnation Family)
○   Ground cover, Edging,
Spring to early summer

Flowers: The ¾- to 1-inch flowers are five-petalled and white. Each petal is notched and appears heart-shaped. The flowers are borne in a cyme with 3 to 15 flowers.

Leaves: Snow-in-summer produces dense mats of 1-inch lanceolate, woolly leaves, which are opposite on the stem. This plant is usually evergreen. The new growth appears very white and woolly; summer leaves usually appear gray-green.

Growing Needs: Snow-in-summer thrives in a full sun location that is dry and even infertile. After flowering, cut back to maintain neatness. Limited self sowing will occur; transplant the small seedlings to create a new ground cover area of snow-in-summer.

Landscape Use:

Height:	6 to 10 inches
Habit:	Low spreading
Spacing:	12 to 18 inches
Texture:	Medium fine

This wonderful addition to the garden is excellent on slopes and draping over walls. It can also be used as an edging and a limited ground cover.

Design Ideas: Snow-in-summer combines well with pink shrub roses. Plant snow-in-summer with grape hyacinths and *Geranium* 'Johnson's Blue' or perennial flax. The early bulbs look nice coming up through or behind it. *Ajuga* 'Burgundy Glow' and snow-in-summer are great edging companions in my garden.

Cultivars:

'Silver Carpet' is more silvery white in leaf and 8 inches tall.
'YoYo' is more compact and is very floriferous.

Problems: None serious.

Propagation: Seed and division.

Zones: 3-7

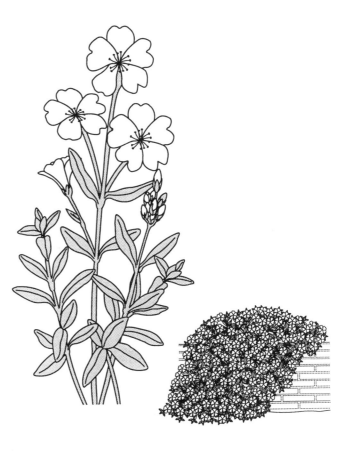

Chrysogonom virginianum

Green and Gold, Goldenstar
Asteraceae (Sunflower Family)
◑, ●   Edging, Spring

Flowers: Each 1½-inch, star-like flower head is yellow. This long-flowering plant has five toothed ray florets with a center of tufted disk florets. Some sparse flowering may occur during the summer.

Leaves: The 1- to 2-inch pubescent leaves are opposite on the stem. Each leaf is ovate with crenate (rounded toothed) margins. Both basal leaves and stem leaves are present.

Growing Needs: Goldenstar does well in partial to full shade but will survive in full sun if there is constant adequate moisture. The soil should be well drained and moist. Add organic matter to maintain a more even soil moisture.

Landscape Use:

Height:	4 to 12 inches
Habit:	Low mounded
Spacing:	12 inches
Texture:	Medium

Goldenstar makes an excellent shady ground cover or edging plant.

Design Ideas: Plant green and gold with heartleaf bergenia, Jacob's ladder, and Bethlehem sage. Add a cheerful spark of yellow to a woodland planting of trillium, Virginia bluebells, common wild ginger, trout lily, and spring beauty.

Cultivars:

'Allen Bush' is 8 inches tall and a fast grower.
'Mark Viette' is 6 inches tall with waxy leaves.

Problems: None serious.

Propagation: Division and self sowing.

Zones: 5-9

Dianthus deltoides

Maiden Pink, Pinks
Carophyllaceae (Pink and Carnation Family)
○, ◑   Edging, Spring to early summer

Flowers: These small, 1-inch flowers are single, solitary (sometimes twos or threes), and come in many shades of reds, pinks, and bicolors, often with a darker ring (eye) in the center. Each five-petalled flower is toothed, making it appear almost feathery. Maiden pink is excellent for long-lasting color in the spring garden. To encourage a second flowering, cut back after the first flush of flowers have faded. Maiden pinks are scentless.

Leaves: The foliage is a low mat of green with sparse, opposite stem leaves. The leaves are 3 to 4 inches long, linear, and narrow.

Growing Needs: Maiden pink performs best in a well-drained, slightly alkaline, moist soil, high in organic matter. It prefers a full sun location, but in a hot summer climate, will appreciate afternoon shade.

Landscape Use:

Height:	4 to 15 inches
Habit:	Low mounded
Spacing:	12 inches
Texture:	Medium fine

Maiden pink will be best suited in the sunny edging or foreground of the spring perennial border. It also makes a nice but limited ground cover or draping plant.

Design Ideas: This plant combines nicely with coral bells and catmint. Try combining maiden pink with evergreen candytuft and grape hyacinth. Pinks and creeping phlox (*P. stolonifera*) bordering a stepping stone path are a welcome spring sight.

Cultivars:

'Albus' has white flowers, 10 inches.

'Brilliant' has bright red flowers.
'Zing Rose' is deep rose pink with a darker ring near the center, 6 inches.

Problems: None serious.

Propagation: Layering, cuttings, division, and seed.

Zones: 4-9

Related Species:

Dianthus plumarius, Cottage Pink, has waxy blue to blue-green foliage (resembling carnation foliage). It is fragrant and usually taller (15 to 18 inches) than maiden pink.

D. barbatus, Sweet William, is a self-sowing biennial that lends charm to the garden with its large rounded cymes of pink, purple, red, white, or bicolors and are good cut flowers. The foliage is bright green and larger than the two mentioned.

Dicentra eximia

Fringed Bleeding Heart
Fumariaceae (Fumitory Family)
◑, ○ (●) Foreground
Spring to summer (fall)

Flowers: These unique flowers are borne on an upright raceme. Each flower is ¾- to 1-inch and narrowly heart-shaped. The lower tips of these pink, red, or white flowers flare out. Trim back faded flowers to encourage reflowering.

Leaves: This plant also offers beautiful blue-green to gray-green foliage. The 8- to 12-inch foliage is compound, finely dissected, and appear almost fern-like. The foliage is attractive until frost.

Growing Needs: Fringed bleeding heart grows best in a partial shade location, but can tolerate a variety of conditions. The soil must be well drained and fertile. A light mulch will benefit this plant.

Landscape Use:

Height:	12 to 18 inches
Habit:	Mounded
Spacing:	12 inches
Texture:	Medium fine

Fringed bleeding heart is lovely as a specimen plant or massed grouping in the shady to moist sunny foreground.

Design Ideas: It looks beautiful with white violets, ferns, and hostas. It also combines well with primroses, forget-me-nots, and Bethlehem sage.

Cultivars:

'Alba' has white flowers.
'King of Hearts' is a compact (8 inches), long-flowering, reddish one.
'Stuart Boothman' is a long-flowering pink with blue-green foliage.

Hybrids:

'Bacchanal' has dark red flowers.
'Luxuriant' has pinkish-red flowers, blue-green foliage, and is long flowering (sparsely in summer).
'Zestful' has pink flowers.
White ones: 'Margery Fish', 'Pearl Drops', 'Snowdrift', 'Snowflakes', *D. formosa* 'Aurora'.

Problems: None serious.

Propagation: Division and seed.

Zones: 3-9

Related Species:

Dicentra spectabilis, Old-fashioned bleeding heart is taller than the fringed bleeding heart. It grows 2 to 3 feet tall and wide. The flowers have a more pronounced heart shape with pink to red outer petals and white tips. The foliage is less dissected and usually dies back in the summer. Plant it next to spreading hostas and ferns.

'Alba' and 'Pantaloons' have white flowers.

Doronicum orientale

(Doronicum cordatum, caucasicum)

Leopard's bane
Asteraceae (Sunflower Family)
◗, ○ Midborder, Spring

Flowers: These showy daisy-like flowers are bright yellow, solitary, 2 to 3 inches wide, and borne high above the mounded foliage.

Leaves: The 2- to 3-inch, light green leaves are cordate (much like that of a violet) and toothed. The basal leaves are long petioled; the stem leaves are alternate and clasp the stem. The foliage usually disappears in summer. Plant Madame Mason, which has more persistent foliage.

Growing Needs: This plant is most appropriately sited in a partially shaded location with moist but well-drained soil. It will tolerate full sun with moist, well-drained soil but does not flower as long. Mulch it!

Landscape Use:

Height:	18 to 24 inches
Habit:	Mounded (foliage)
	two tiered (in flower)
Spacing:	12 inches
Texture:	Medium

Leopard's bane makes a great show in the shady midborder of the garden.

Design Ideas: This cheerful spring plant looks great with bleeding hearts and epimediums. Plant a hosta or ferns nearby to mask the fading foliage of both old-fashioned bleeding heart and leopard's bane. Also, try combining it with 'Cheerfulness' daffodils, perennial forget-me-nots (*Brunnera*), and sweet woodruff.

Cultivars:

'Madame Mason' ('Miss Mason') has more persistent foliage.

'Magnificum' has larger flowers and is 2 to 2½ feet tall.

'Spring Beauty' has a double flower and is 15 to 18 inches tall.

Problems: None serious.

Propagation: Seed and division in summer after flowering.

Zones: 4-7

Epimedium × rubrum

(E. alpinum var. rubrum)

Red Alpine Epimedium, Barrenwort,
  Bishop's Hat
Berberidaceae (Barberry Family)
◑, ●   Ground cover, Edging, Spring

Flowers: Red alpine epimedium produces airy racemes of graceful ¾- to 1-inch, columbine-like flowers. Each flower has crimson inner sepals, flushed with yellow or red, and is surrounded by four short, backward projecting, creamy yellow spurs. The loose, airy racemes appear before or just as the leaves emerge.

Leaves: Each new heart-shaped leaf emerges red and then turns green often displaying a red edge. The basal leaves arise on long thin petioles and are biternate with a spiny, serrate margin. In the fall, the leaves turn red and then brown in winter. The persistent winter foliage can be mowed (large areas) or removed in early spring before the flowers appear. An excellent four season plant!

Growing Needs: This excellent shade plant will grow well in both part to dense shade with adequate moisture and mulching to maintain a healthy plant. The soil should be well drained. Barrenwort tolerates dry shade and competition from tree roots. Cut back the old foliage in March. It will not require division for many years.

Landscape Use:

Height:	12 inches
Habit:	Mounded
Spacing:	12 inches
Texture:	Medium to medium fine

Red alpine epimedium makes an excellent shady ground cover or edging. Plant it as a showy single specimen. It performs well under trees and looks great under flowering crab apples and redbuds as well as denser shade trees.

Design Ideas: Red alpine epimedium combines well with wild blue phlox and daffodils.

White wood hyacinths and heartleaf bergenia make a nice backdrop for red alpine epimedium.

Problems: None serious.

Propagation: Division of rhizomes.

Zones: 4-8

Related Species:

E. grandiflorum, 'Longspur Epimedium' has white, pink, red, or purple flowers (cultivars) and is 12 to 18 inches tall.

E. × perralchicum 'Frohnleiten' is a vigorous 12 to 18 inches epimedium with yellow flowers prominently borne above the newly emerging reddish (and toothed) foliage. Foliage is evergreen.

   'Wisley' is taller (1½ feet) with bolder evergreen foliage and yellow flowers.

E. × versicolor 'Sulphureum' has pale yellow flowers and tolerates drier, sunnier sites, 8 to 12 inches.

E. × warleyense has orange flowers and a height range of 1 to 2 feet.

E. youngianum 'Niveum' is a 8- to 12-inch, white, flowering ground cover.

   'Roseum' has pinkish-mauve flowers.

Galium odoratum

(Asperula odorata)

Sweet Woodruff
Rubiaceae (Madder Family)
◑, ●, ○   Ground cover, Edging, Spring

Flowers: Sweet woodruff produces small, ¼-inch white, star-like (four-parted) flowers in cymes.

Leaves: Six to eight 1-inch leaves are whorled around the four-angled stem. Each leaf is narrowly ovate to lanceolate, bristle-tipped, and finely toothed along the margins. To notice a fragrance, crush the leaves and stems for a spicy new mown hay smell.

Growing Needs: Deep to partial shade are the best locations for this plant; full sun with moisture will also work. A well-drained, slightly acidic soil with high organic matter is preferred. Sweet woodruff is a vigorous plant in areas with plenty of moisture.

Landscape Use:

Height:	6 to 8 inches
Habit:	Spreading
Spacing:	12 to 18 inches
Texture:	Medium fine

Sweet woodruff is a useful ground cover for shade or sun. If used as an edging plant, give it some room to grow—it may be too vigorous for a small area.

Design Ideas: Sweet woodruff looks nice with Bethlehem sage and columbines. Zumi crabapple and sweet woodruff make a nice duet since the flowering times are the same. Plant pansies near *Galium odoratum* with daffodils and tulips behind them for a cheerful spring grouping. For a sunny spot, plant pinks, *Coreopsis rosea*, purple coneflower, and Powis Castle artemisia near it for a pink, white, and silver combination. Sweet wood-ruff is a trooper in the garden and will mask bare spots of old-fashioned bleeding heart and bulbs.

Problems: None serious.

Propagation: Division almost anytime during the growing season.

Zones: 4-8

Geranium sanguineum

Bloodred Cranesbill (Crane's Bill) or Hardy
  Geranium
Geraniaceae (Geranium Family)
○, ◑ Edging or foreground, Spring to summer

Flowers: The five-petalled flowers are 1½
inches wide and borne in a terminal cluster. Flower
colors range from magenta for the straight species
to white, pink, and blue for the cultivars or varieties. The fruit of this plant has a protruding "beak,"
resembling the bill of a crane, hence the common
name, cranesbill.

Leaves: The 2- to 3-inch leaves are rounded in
outline with five to seven slender lobes. The leaf has
some scattered surface pubescence; the stem and
flower peduncles are covered with long white hairs.
This plant has lovely foliage throughout the growing season and turns red to burgundy in the fall.

Growing Needs: Any full sun to partial shade
location is suitable for bloodred cranesbill. A well-
drained soil of moderate fertility is suitable. Avoid
moist sites of high fertility or organic matter because the straight species will spread very quickly.
The cultivars/varieties are less vigorous.

Landscape Use:

Height:	12 to 18 inches
Habit:	Mounded
Spacing:	18 inches
Texture:	Medium

Bloodred cranesbill will work well as an edging to
foreground plant in the perennial border. All of
the hardy geraniums are good blending plants in
the garden. The straight species can be used as a
ground cover in dry and shady spots or on slopes.

Design Ideas: All of the hardy geraniums
blend well with spring bulbs and iris behind them.
Sea pink (*Armeria*) and snow-in-summer are good
edging plants in front of the hardy geraniums. A
particularly nice grouping with *G. sanguineum*
'Alba' is soapwort (*Saponaria ocymoides*), perennial
flax, and foxglove. The leaves of fall anemone and
hardy geraniums enhance each other as well as having two separate seasons of flowering.

Cultivars:

'Album' has white flowers.

'Alpenglow' has bright fuchsia flowers, is long flowering, 8 inches.

var. striatum ('Lancastriense') has lovely pale pink flowers with deeper pink veins and is an 8-inch long bloomer.

var. striatum 'Purple Flame' has bright reddish-purple flowers, finely cut foliage, and a 6- to 8-inch mounded habit.

Problems: Leaf spots.

Propagation: Seed and division.

Zones: 3-8

Every garden must have at least one!

Related Species:

Geranium × 'Johnson's Blue' has a mounded to sprawling habit and nice blue flowers.

G. cantabrigiense 'Biokova' has beautiful soft leaves (red in fall) and white flowers with a pink blush, 10 inches.

'Cambridge' has a pink flower, 10 inches.

'Karmina' has striking carmine rose flowers and is a spreading 10 inches.

G. cinereum 'Ballerina' is 6 inches high with lilac-pink, veined with purple, flowers and deeply lobed green leaves.

G. clarkei 'Kashmir White' has numerous beautiful white flowers above finely dissected leaves, 12 inches.

'Kashmir Blue', 'Kashmir Pink', 'Kashmir Purple'.

G. dalmaticum is a beautiful dwarf geranium at 6 inches with very small leaves (½ inch) and pink flowers (1 inch). The orange fall color is distinctive.

G. endressii 'Wargrave Pink' has pink flowers, grayish-green foliage, is 15 to 18 inches, and long flowering.

G. himalayense 'Lilac Cranesbill' is 18 to 24 inches with purple flowers and large deeply divided foliage.

'Gravetye' has intense blue flowers with reddish-purple centers and veins and orange-red fall color, 15 inches.

'Plenum' ('Birch Double') has orchid purple, doubled flowers, 15 inches.

G. macrorrhizum 'Bigroot Geranium' is aromatic in leaf, which are five- to seven-lobed. Flowers are pink, 1 to 1½ feet.

'Album' is white with red sepals.

'Bevan's Variety' is a compact 12 inches with magenta flowers.

'Ingwersen's Variety' has lovely pubescent leaves that turn red in the fall and light pink flowers.

'Variegatum' has cream variegated leaves and pinkish purple flowers.

G. maculatum 'Wild Geranium' is a lavender-pink native plant with nicely divided foliage, 15 to 24 inches.

'Alba' is the white form.

G. × magnificum has violet-blue flowers and rounded leaves, 2 feet.

G. × oxonianum 'Claridge Druce' has purplish-pink flowers with darker venation above leathery grayish-green foliage, 18 inches to 2 feet.

'Rose Clair' has silvery pink flowers with purple veins, turning bluish-white with age, 15 to 18 inches.

G. phaem 'Mourning Widow' has deep maroon flowers, lobed foliage, and a loose habit, 1½ to 2½ feet.

G. platypetalum has violet-blue flowers that are saucer-shaped. The foliage is spicy scented, 18 inches.

G. pratense 'Mrs. Kendall Clarke' has lovely sky blue flowers and lobed foliage, 2 feet.

G. psilostemon 'Armenian Cranesbill' has brilliant magenta flowers with a black eye and veins, large-lobed green leaves, 2½ feet.

'Bressingham Flair' has a lilac flower with a dark eye.

G. × 'Ann Folkard' has purplish magenta flowers and yellowish-green leaves. This group may need peastaking.

G. renardii has unique, thick, leathery, rounded, wrinkly green leaves with blue-veined white spring flowers, 12 inches.

G. sylvaticum 'Mayflower', Wood Cranesbill, is 3 feet with deep blue, white-centered flowers and lobed foliage.

Related Genus:

Erodium glandulosum Storksbill, Heronsbill, is a hardy geranium cousin. It has long-flowering small lilac flowers with darker venation borne in open umbels and bipinnate pubescent basal foliage. The well-mulched plant has overwintered in my garden (zone 5) for four years although it is considered a zone 6 plant.

Iris × germanica (rhizomatous hybrids)

Bearded Iris, Garden Iris, Poor Man's Orchid
Iridaceae (Iris Family)
○ Foreground, Midborder, Spring to early summer

Flowers: This long-standing garden favorite has a 2- to 4-inch flower composed of three standards and three falls. The falls have a crest of hairs that forms the "beard." The bearded iris comes in many colors—white, blue, yellow, purple, brown, pink, and bicolors.

The six horticultural divisions of Iris each have a slightly different flowering time and height as follows:

Miniature Dwarf (*pumila*) — 4 to 9 inches, April.

Standard Dwarf — 10 to 15 inches, early May.

Intermediate — 15 to 28 inches, mid May.

Miniature Tall — 18 to 26 inches, late May.

Border — 28 inches, late May.

Standard Tall — 28 inches, late May.

Leaves: The leaves are 6 to 10 inches or longer, arise from the rhizome, and are arranged in a fan shape. The leaves are lanceolate, upright, and gray to blue green.

Growing Needs: Bearded iris does best in a full-sun, well-drained area. Plant the rhizomes at the soil surface with the roots just below the surface. To avoid iris borer damage, practice prevention and cleanliness in the garden. Remove and destroy spent iris leaves in the late fall and regularly divide and discard the old rhizomes.

Landscape Use:

Height:	4 inches to 3 feet
Habit:	Vase shaped
Spacing:	18 inches
Texture:	Medium to bold

Plant bearded iris in the sunny foreground or mid-border as a specimen plant or massed grouping. Bearded iris makes an excellent addition for the cut flower garden.

Design Ideas: Bearded iris looks great with columbines, poppies, peonies, and blue indigo. The purple and white bi-colored Stepping Out iris looks great with pink *Centranthus* and Silver King artemisia.

Cultivars: Numerous!

Problems: Iris borer, followed by bacterial soft rot.

Propagation: Division every four to five years in June or July after flowering. Cut the foliage back to a 4- to 6-inch fan before replanting.

Zones: 3-10

Related Species:

Iris pallida is a 15- to 20-inch iris with spring lavender flowers. 'Argentea Variegata' blue-green leaves with white variegation, 'Aurea Variegata' yellow variegation.

Lamium maculatum cultivars

Lamium, Dead Nettle
Lamiaceae (Mint Family)
◑, ● Ground cover, Edging, Late spring to summer

Flowers: The 1-inch, rose-pink or white, hooded and lipped flowers are borne in whorls on erect stems.

Leaves: This wonderful foliage plant has silver mottled or striped 1 to 3 inch leaves. Each leaf is crinkled, oval to triangular, and opposite.

Growing Needs: Lamium does well in a partially to densely shaded location. The soil should be well drained although lamium will tolerate both dry to wet conditions. The cultivars are preferable due to the vigorous nature of the species.

Landscape Use:

Height:	10 to 12 inches
Habit:	Mounded
Spacing:	18 inches
Texture:	Medium

Dead nettle is a good shady ground cover to follow after the spring bulbs have flowered. Use as a bright edging plant in shade.

Design Ideas: Interplant spring bulbs with lamium along with creeping lilyturf, *Liriope. Spicata* Beacon Silver lamium combines nicely with *Heuchera* 'Palace Purple', pink *Phlox stolonifera* (creeping phlox), and violas or pansies in whites or pinks. Use the golden-leaved ones as charming accents or ground covers. White Nancy contrasts well with *Brunnera* (spring forget-me-not).

Cultivars:

'Aureum' has lavender-pink flowers with yellowish-green leaves, 6 inches.

'Beacon Silver' has pink flowers, distinctive silvery leaves with a narrow green edge, 4 to 8 inches.

'Beedham's White' has bright yellowish-green foliage with some white variegation and white flowers.

'Chequers' has deep pink flowers and silvery variegated foliage.

'Pink Pewter' has silvery variegated leaves and pale pink flowers.

'Shell Pink' is a long-flowering pink.

'White Nancy' is similar to 'Beacon Silver' but has white flowers.

Problems: Slugs.

Propagation: Division and cuttings.

Zones: 3-8

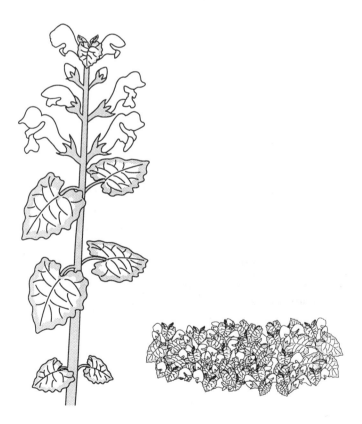

Paeonia lactiflora

Peony
Ranunculaceae (Buttercup Family)
○ Background, Late spring

Flowers: These large, 3- to 8-inch, fragrant flowers come in several forms: single, Japanese, semi-double, double, and bomb. Peonies come in many different colors including red, pink, peach, purple, yellow, white, and bicolors. Longer flowering can be achieved by selecting early, mid-season, or late cultivars.

Leaves: The alternate and compound leaves are 6 to 12 inches and biternate, divided into three leaflets that are each lanceolate and lobed. When they first emerge, they are a beautiful red color. The shiny foliage remains attractive all through the growing season. Do not cut foliage back until fall or winter.

Growing Needs: Peonies grow well in full sun but will tolerate light shade, especially the pastel colors. The soil should be well drained with moderate fertility and neutral to slightly alkaline pH. Deeply prepare the soil to a 12-inch depth. The side buds can be disbudded (leaving only the top bud) to remove weight from the plant and produce one large, special flower. Or, leave all the buds to prolong the flowering time. Some varieties may need to be peastaked, caged, or staked. Remove faded flowers. Transplanting is rarely needed. However, if moving a peony, plant a division that has three to five eyes during August or early September. Plant the eyes (the reddish buds showing on the roots) 1 inch deep.

Landscape Use:

Height: 2 to 3 feet
Habit: Mounded
Spacing: 3 feet
Texture: Medium to medium bold

Peonies are excellent for the background as a small mass or hedge or as a showy single specimen. Peonies are excellent cut flowers and will dry beautifully. Harvest them at their peak.

Design Ideas: The traditional favorite is placing bearded iris and peony together. Also, peony combines nicely with columbines, Siberian iris, painted daisy, and edgings, such as snow-in-summer, pinks or *Geranium sanguineum* 'Alba'.

Problems: Botrytis, Phytophthora.

The spread of these problems can be mostly prevented by sanitation. Remove and destroy any infected flower or foliage as it is seen, as well as all of the foliage before the winter.

Not a Problem: Ants! Ants are attracted to the nectar of the peony and are not harmful to the plant. Brush the ants off before bringing the flowers inside as a cut flower. Or dip the flower in a bucket of water to remove the ants.

Propagation: Division to three to five eyes (buds) and replant 1 inch deep.

Zones: 3-8

Related Species:

Paeonia tenuifolia, Fernleaf Peony, has finely divided foliage, dark red flowers, and is 18 to 24 inches.

Paeonia suffructicosa, Tree Peony has woody stems (should not be cut back in fall) and a beautiful range of lovely large 6- to 8-inch flowers.

Papaver orientale

Oriental Poppy
Papaveraceae (Poppy Family)
○ Midborder, Late spring to early summer

Flowers: Each 4- to 7-inch flower has four or six petals and are usually deep scarlet red, salmon, orange, pink, or white with black markings inside the petals and showy black stamens. These petals are thin and papery and emerge from the bristly bud very wrinkled and crinkled. The flowers are borne on long bristly peduncles. The poppies seed pods persist and are distinctive.

Leaves: Oriental poppy leaves look thistle-like when emerging. The bristly (hispid), toothed leaves are large, 10 to 12 inches long, and pinnately divided. The leaves are borne both in a basal rosette and alternate on the stem.

Growing Needs: A full sun, well-drained site is best. Poppy foliage will die back in midsummer. The foliage will then emerge in fall and over-winter. Some taller varieties of poppies will need to be staked or placed near supporting plants.

Landscape Use:

Height:	2 to 3 feet
Habit:	Mounded
Spacing:	18 to 24 inches
Texture:	Medium bold

This beautiful flower makes an excellent specimen plant. Or plant oriental poppy in the midborder of the garden.

Design Ideas: Baby's breath is an excellent combination with oriental poppy for softening the poppy's boldness and masking its lack of summer foliage. The lemon yellow *Achillea* 'Moonshine' compliments a scarlet red or orange poppy grouping. Poppies and purple or white bearded iris combine well with nicely contrasting foliage.

Cultivars:

Reds — 'Brilliant', 'Indian Chief', 'Ladybird', 'Warlord'.

Orange Reds — 'Allegro', 'Glowing Embers', 'Nana Flore Pleno' (ruffled).

Orange— 'Prince of Orange'.

White— 'Carousel' (orange edge), 'Fatima' (pink edge), 'Perry White'.

Pink or Salmon — 'Cedric Morris', 'Helen Elizabeth', 'Maiden's Blush', 'Mrs. Perry', Princess Victoria Louise', 'Raspberry Queen'.

Problems: None serious.

Propagation: Seed (save your own), division, and root division.

Zones: 3-8

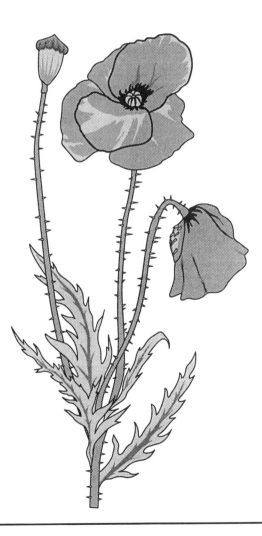

Polemonium caeruleum

Jacob's Ladder
Polemoniaceae (Phlox Family)
◑ Foreground, Midborder, Spring to early summer

Flowers: Jacob's ladder has ½- to 1-inch clusters of cupped, nodding, sky blue, white, or lavender flowers.

Leaves: The 5- to 6-inch, apple-green leaves are alternate on the stem and are pinnately compound with 1- to 1½-inch lanceolate leaflets. The numerous leaflets are symmetrically arranged on the leaf to appear like a ladder. This foliage appears very clean and neat and is attractive all growing season.

Growing Needs: A partial shade location is well suited for this plant. Plant it in a moist but well-drained site with added organic matter.

Landscape Use:

Height:	15 to 24 inches
Habit:	Upright to mounded
Spacing:	12 to 18 inches
Texture:	Medium fine

Jacob's ladder makes an excellent shady foreground to midborder plant.

It looks nice massed or as a small accent planting. It is also good for naturalizing in an area.

Design Ideas: Group the blue Jacob's ladder with other moisture-loving plants, such as yellow globeflower and white astilbe with an edging of yellow barren strawberry. Jacob's ladder combines well with many wild flowers, such as Jack in the pulpit, Virginia bluebells, ferns, and bleeding hearts.

Cultivars:

'Album' has a white flower.

'Apricot Delight' is lilac-tinged with apricot and flowers profusely, 20 inches.

Problems: None serious.

Propagation: Seed, division, and stem cuttings.

Zones: 3-7

Related Species:

P. foliosissimum has long-flowering, lilac blue flowers and finer foliage than *P. caeruleum,* 20 to 24 inches.

P. reptans 'Blue Pearl' has a light blue flower.

Thermopsis villosa

(caroliniana)

Carolina Lupine
Fabaceae (Legume Family)
○, ◑ Background, Late spring to early summer

Flowers: The ¾-inch, yellow, pea-like flowers are borne on 8 to 10 inch racemes. The flowers resemble a yellow lupine (and also blue indigo).

Leaves: The light green leaves are alternate and palmately compound. Each leaflet of three has a pair of leafy stipules at the end of its long petiole.

The leaflets are ovate to obovate, 2 to 3 inches long, and pubescent on top and glaucous underneath. This attractive foliage provides a good garden background or filler all through the growing season.

Growing Needs: Carolina lupine prefers a full sun, well-drained site but will tolerate a partially shaded one. Once planted and established, Carolina lupine is very low maintenance, requiring little or no fertilizer or supplemental watering. Place it where the soil can be deeply prepared before planting. Because of its taproot, it is very drought resistant but this also makes it hard to transplant.

Landscape Use:

Height: 3 to 4 feet
Habit: Upright mounded
Spacing: 2 feet
Texture: Medium

Carolina lupine makes an attractive sunny background plant in the garden. It is also a nice addition to a cut-flower garden.

Design Ideas: Carolina lupine looks great as the background to perennial blue salvia (midborder) with yellow corydalis as an edging. For late spring to early summer interest, I have placed the yellow *Thermopsis* and the blue *Baptisia australis* (blue indigo) next to each other in the center of an island bed next to feather reed grass (*Calamagrostis*) and Russian sage. Fall asters add autumn interest with summer-flowering blue balloon flowers, gooseneck loosestrife, and yellow daylilies. The yellow, blue, and white color scheme is quite pleasing.

Problems: None serious.

Propagation: Seed, transplanting of young seedlings (rarely self sows). Purchase and plant small container plants of Carolina lupine.

Zones: 3-8

Trollius europaeus

Globeflower
Ranunculaceae (Buttercup Family)
◑, ○   Foreground, Midborder, Spring to early summer

Flowers: These flowers are solitary or with two on a stem. Each 1- to 2-inch, globe-like or cupped-shaped flower is made up of many yellow (lemon or golden) petaloid sepals.

Leaves: The basal leaves are palmately divided into three to five lobes and have long petioles. The stem leaves are three-lobed and sessile (no petiole). Each dark green leaf is 4 to 6 inches and toothed.

Growing Needs: This is an excellent plant for moist soils. Partial shade to sun is the best site for globeflower. In lighter soils, add organic matter and mulch. Remove faded flowers for a sparse re-bloom.

Landscape Use:

Height:	1½ to 2 feet
Habit:	Mounded
Spacing:	18 inches
Texture:	Medium

Globeflower is wonderful planted in the foreground to midborder near water or boggy sites.

Design Ideas: Globeflower looks lovely with Japanese iris, especially in or around a water garden. Ferns and wildflowers combine well with globeflower. Combine Jacob's ladder, astilbe, and candelabra primroses with globeflower for a lovely spring effect.

Cultivars:

'Grandiflorus' is a large bright yellow.
'Superbus' is a late-flowering yellow.
'Verdus' has a lemon yellow flower.

Problems: Powdery mildew.

Propagation: Seed and division.

Zones: 3-8

Related Species

T. × 'Cheddar' is a very pale yellow and grows to 28 inches.

T. × *cultorum* 'Lemon Queen' (20 inches) and 'Golden Queen' (30 to 36 inches) are lemon yellow and golden yellow respectively.

'Alabaster' is a creamy white, 2 feet.

'Earliest of All' is an early golden yellow.

'Orange Globe' is a tall (36 inches) orange globe-flower.

Waldsteinia ternata

Barren Strawberry
Rosaceae (Rose Family)
○, ◑   Edging, Spring to early summer

Flowers: Each 1-inch, bright yellow flower is five petalled. There are three to eight flowers borne in a corymb. The fruit of barren strawberry is small and inedible.

Leaves: The evergreen leaves are similar to strawberry (*Fragraria*). Each 2- to 3-inch leaf is trifoliate and pubescent. The petioled leaves arise from a basal rosette.

Growing Needs: A partial shade to sun location will do. The soil should be well drained and moist especially in full sun; mulching is beneficial to retain moisture. Dry sites will cause unsightly browning of the leaf edges. Barren strawberry will spread easily by runners.

Landscape Use:

Height:	6 to 8 inches
Habit:	Low mounded
Spacing:	8 to 12 inches
Texture:	Medium

Barren strawberry makes a nice edging plant or a limited ground cover.

Design Ideas: Combine barren strawberry with columbines and foxgloves. In a woodland setting, it makes a nice edging for wildflowers and bulbs as well as primroses and hostas.

Problems: None serious.

Propagation: Division.

Zones: 4-7

Related Species:

W. fragrarioides is very similar to *W. ternata* but is more clump-forming.

Related Genus:

Fragaria 'Pink Panda' and 'Lipstick' are pink, ornamental, 6- to 8-inch spreading ground covers.

Other Spring Perennials

Anemone sylvestris — **Snowdrop Anemone**

Full sun to partial shade.

White flowers.

Edging to foreground, 12 to 18 inches.

Mounded, easy to grow.

Arum italicum 'Pictum' — **Painted Arum**

Partial to dense shade.

Grown for orange-red berries.

Attractive veined foliage, 12 to 18 inches.

Astilboides tabularis — **Rodger's Flower**

Partial shade, moist.

White panicles.

Large rounded peltate leaves.

Mounded, two tiered, 3 to 4 feet.

Background, near water.

Astrantia major — **Masterwort**

Partial shade.

Pink, red, white, and unique.

Foreground to midborder.

Two-tiered habit, 2 to 3 feet.

Corydalis lutea — **Yellow Corydalis**

Sun, partial shade, well drained.

Yellow racemes, resembling *Dicentra*.

Fern-like foliage.

Mounded, 1 foot, foreground.

Crambe cordifolia — **Colewort**

Full sun, background.

Showy, large white panicled flowers, resembling a large baby's breath.

Two tiered or mounded.

5 to 6 feet tall, 3 to 4 feet wide.

Filipendula vulgaris — **Dropwort Meadowsweet**

Sun to partial shade.

White panicles.

Midborder, 2 to 3 feet, two tiered.

Geum hybrids — **Geum, Avens**

Full sun to partial shade.

Red, orange, yellow.

Foreground to midborder.

Two tiered, 18 to 24 inches.

Gillenia trifoliata — **Bowman's Root**

Partial shade, sun, moist, well-drained.

White starry flowers in panicles.

Trifoliate (three-part) leaves, red in fall.

Mounded, 1 to 3 feet, foreground.

Hesperis matronalis — **Dame's (or Sweet) Rocket**

Partial shade to full sun.

White or lavender (phlox-like).

Midborder to background, 3 to 4 feet.

Upright habit, self-sowing biennial.

Iris pseudoacorus — **Yellow Flag (Iris)**

Partial shade, full sun, moist.

Yellow iris (resembles Siberian iris)

Upright, 5 feet, edge-of-pond.

Leontopodium alpinum — **Edelweiss**

Fuzzy, white, star-like flowers.

Sunny edging, 6 to 10 inches.

Linaria sp. —

Full sun, well-drained site, miniature snapdragon-look.

L. alpina — **Alpine Toadflax**

Purple with yellow, 6 inches, mounded edging.

L. genistifolia var. *dalmatica* — **Toadflax**

Long-flowering, yellow, not rambunctious.

Linaria vulgaris Butter and Eggs — Upright, 24 to 30 inches, midborder, self sows.

L. purpurea — **Purple Toadflax**

2 to 3 feet.

'Alba' — white, 'Canon J. Went' — pink.

Upright, midborder.

Lysimachia nummularia — **Moneywort**

Sun to shade, yellow flowers.

Round leaves.

Low spreading, 2 to 4 inches.

'Aurea' — yellow foliage, less invasive.

Lysimachia punctata — **Yellow Loosestrife**

Sun or partial shade.

Yellow, upright mounded.

Old-fashioned plant, 2 to 3 feet.

Can be invasive.

Myosotis sylvatica — **Forget-Me-Not**

Partial shade to sun, edging.

Blue racemes (pink, white).

Mounded, 6 to 8 inches.

Self-sowing biennial

Potentilla nepalensis — **Cinquefoil**

Full sun, well drained.

Pink, five-petalled flowers.

Compound leaves.

Cultivars 12 to 15 inches.

Mounded, foreground.

Thalictrum aquilegifolium — **Columbine Meadow Rue**

Partial shade.

Pink to mauve fluffy flowers.

Midborder to background, 2 to 3 feet.

Tiarella cordifolia (stoloniferous)
T. wherryi (clump-forming) — **Foam Flower**

Partial to dense shade, moist.

Thin fluffy white racemes.

Attractive toothed foliage (resembling coral bells), two tiered, 6 to 12 inches.

Edging, ground cover.

Tradescantia × *andersoniana* — **Spiderwort**

Partial shade or sun.

Purple, rose-violet, white, blue, red, pink flowers, three-petalled in terminal umbels.

Foreground, 16 to 24 inches.

SPRING BULBS

Allium moly

Lily Leek, Golden Garlic
Liliaceae (Lily Family)
○   Edging, Spring to early summer

Flowers: Lily leek produces umbels with glorious, ¼-inch, yellow star-shaped flowers. Very attractive!

Leaves: The two basal leaves are medium green, waxy, and 10 to 12 inches long. Each leaf tapers at both ends and is 1 inch wide.

Growing Needs: This plant should be sited in the full sun. A well-drained soil is best. Very easy to grow!

Landscape Use:

Height:	8 to 12 inches
Planting depth:	4 inches
Spacing:	4 to 6 inches
Planting Time:	Fall
Texture:	Medium fine

Lily leek deserves its place in the garden. Interplant it with low ground covers or place it in the foreground of the border.

Design Ideas: Lily leek looks great with red late-season tulips and evergreen candytuft. Combine this plant with snow-in-summer and *Ajuga* 'Pink Beauty'. Masses of cheerful lily leek combine well with the silver leaves of artemisias or lamium. The groups fill in and look nicer each year.

Problems: None serious.

Propagation: Division in fall and self sowing.

Zones: 4-8

Related Species:

Allium neapolitanum, White Allium, produces a white flower and can grow to 24 inches.

Allium ostrowskianum, Red Allium, resembles lily leek but produces a red flower.

Convallaria majalis

Lily of the Valley
Liliaceae (Lily Family)
◑, ● (○)　Edging, Spring

Flowers: These fragrant, ¼-inch, waxy, bell-shaped flowers are produced along a one-sided raceme. Each flower is white or light pink.

Leaves: The ovate to lanceolate leaves are glabrous and frame the flowers. Each leaf is 6 to 8 inches long and 2 to 3 inches wide.

Growing Needs: A well-drained site is preferable. These easy-to-grow plants are most prolific in the partial shade, but will grow in deeper shade with less flowering. In dry full sun, the leaves become unsightly in the summer. Add organic matter to add moisture-holding capability to the soil. Lily of the valley are grown from horizontal rhizomes and pips. Pips are the upright buds or shoots arising from the rhizome.

Landscape Use:

Height:	6 inches
Planting Depth:	Just below the surface.
Spacing:	6 inches
Texture:	Medium (flower)
	Medium bold (leaf)

Lily of the valley makes an excellent shady ground cover or edging plant. It is great for adding fragrance to the garden.

It is also a popular and (once again) trendy fresh cut flower. Lily of the valley is quite at home in an old-fashioned flower border.

Design Ideas: Combine lily of the valley with Bethlehem sage and foam flower, with astilbe behind them and European wild ginger in front. Lily of the valley and hostas go well together. Ferns are a good choice to soften the bolder textures of those two.

Cultivars:

'Aureo-variegata' ('Striata') has white variegation (stripes) on its green leaves.

'Flore Pleno' has larger, double, white flowers.

'Fortin's Giant' is larger in size and flowers and is a good strain for forcing the flowers into early bloom. Lift, re-pot, and give them a warm window.

'Rosea' has pale pink flowers.

Problems: None serious.

Propagation: Division. The pips can be divided in early spring or after flowering if overcrowded. Otherwise, frequent division is not necessary except to propagate it.

Zones: 3-8

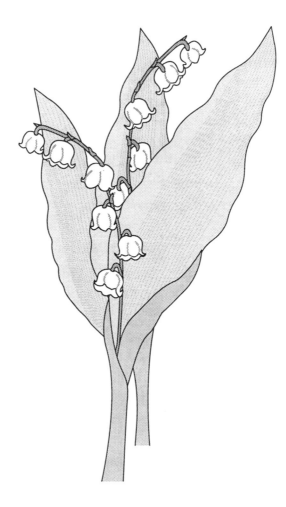

Hyacinthoides hispanica

(Endymion hispanica, Scilla campanulata, Scilla hispanica)

Wood Hyacinth, Spanish Bluebell, Spanish Squill
Liliaceae (Lily Family)
◑, ○   Foreground, Late spring to early summer

Flowers: Wood hyacinth produces delicate nodding bells of ¾-inch, blue to lavender, white, or pink flowers on a raceme. It flowers later than most tulips and daffodils therefore making it a good transition plant before bearded iris.

Leaves: The attractive 12-inch leaves form narrow, linear, vase-shaped clumps. The medium green foliage will eventually die back in midsummer.

Growing Needs: Partially shaded planting sites are best, although it will grow in the sun. The soil should be well drained.

Landscape Use:

Height:	10 to 15 inches
Planting Depth:	2 to 3 inches
Spacing:	4 to 6 inches
Planting Time:	Fall
Texture:	Medium to medium fine

Wood hyacinth makes an excellent grouping in the foreground of the garden. Or, use it interplanted with ground covers.

Design Ideas: Wood hyacinth is beautiful in English ivy. It makes a wonderful combination with European wild ginger, heartleaf bergenia, and white violets. White wood hyacinths blend well with the blue wild sweet William (*Phlox divaricata*); the phlox foliage masks the fading summer foliage of wood hyacinth.

Cultivars:

(May be hybrids of *H. hispanica* × *H. non-scripta*)
'Alba' has white flowers.
'Danube' has prolific deep blue flowers.
'Excelsior' is tall and has large blue violet flowers each with a blue stripe.
'Queen of the Pinks' has deep pink flowers.
'Rose' has pink flowers.

Problems: None serious.

Propagation: Division in fall.

Zones: 4-9

Ornithogalum umbellatum

Star of Bethlehem
Liliaceae (Lily Family)
○, ◑, ●   Foreground, Late spring

Flowers: Each ½- to 1-inch, star-shaped flower has six petals, which are borne in a broad corymbose raceme. The three outer petals have a green stripe.

Leaves: The 10-inch, basal, strap-like leaves are green and linear. Clumps look like tufts of grass in the garden.

Growing Needs: Star of Bethlehem will grow in either shade or sun. The soil should be well drained. This plant multiplies fast, so beware.

Landscape Use:

Height:	9 to 12 inches
Planting depth:	3 inches
Spacing:	6 inches
Planting Time:	Fall
Texture:	Medium fine

Star of Bethlehem can be a garden bully, edging out less assertive plants, so choose your placement of it well. It will quickly take over a civilized perennial garden. Plant it in a confined area, bordered by sidewalks and the foundation of your home. It fits nicely into an informal naturalized area.

Design Ideas: Star of Bethlehem looks nice interplanted in, around, and under a shrub border. Place it with ornamental grasses. Use it in a raised bed area with *Ajuga* and *Waldsteinia* (barren strawberry), that is, other plants that will stand up to it and not let it over-take them.

Problems: None serious.

Propagation: Seed and division.

Zones: 4-8

Other Spring Bulbs

Allium aflatunense — Persian Onion

Full sun, well drained.

Round umbel of starry lavender flowers, strappy foliage.

Upright, 3 feet, midborder.

Camassia leichtinii (C. quamash) — Quamash, Camass

3 to 4 feet. (C. quamash — 2 to 2½ feet.)

Full sun, partial shade, well drained.

Eye-catching racemes of blue or white (lavender) starry flowers.

Grass-like foliage, two tiered, vase shaped.

Midborder to background

Under-used and lovely!

Fritillaria imperialis — Crown Imperial

Partial shade, moist, fertile, humus-rich, not long-lived.

Unique red-orange, orange, yellow.

Clusters of nodding bells with green tufts above. Unique musky odor.

Two-tiered habit, background or center of an island bed, 3 to 4 feet. 'Aurora' is bright orange-red.

'Lutea Maxima' has large lemon yellow flowers.

'Premier' is large and soft orange.

'Rubra Maxima' has brownish-orange flowers, shaded with red.

Fritillaria persica — Persian Fritillary

Partial shade, moist, well drained.

Somber dark purple racemes.

Upright, 2 to 3 feet, background, massed.

Fritillaria meleagris — Checkered Lily, Guinea-Hen Flower, Snake's Head Fritillary

Partial shade to full sun, moist, well-drained soil.

Checkerboard mix of purple, gray, and ivory or white. The drooping bells actually look checkered.

Vase-shaped clump as an accent grouping or massed in the foreground, 10 to 15 inches.

'Alba' is a lovely white form.

Achillea filipendulina

Achillea ptarmica

Aconitum napellus

Ajuga reptans

Alchemilla mollis

Allium moly

Allium giganteum

Amsonia tabernaemontana

Anchusa azurea

Anemone × hybrida

Aquilegia hybrid

Arabis caucasica

Armeria maritima 'Alba'

Artemisia schmidtiana 'Silver Mound'

Asarum europaeum

Asclepias tuberosa

Aster alpinus

Astilbe × arendsii (center)

Aubrieta deltoidea

Baptisia australis (left background)

Belamcanda chinensis

Bergenia cordifolia

Brunnera macrophylla

Campanula carpatica

Campanula glomerata

Centaurea macrocephala

Centaurea montana

Centranthus ruber

Cerastium tomentosum

Ceratostigma plumbaginoides

Chionodoxa luciliae

Chrysanthemum × morifolium

Convallaria majalis

Chrysogonom virginianum

Coreopsis grandiflora

Coreopsis verticillata (center)

Crocus chrysanthus

Dianthus deltoides

Delphinium elatum

Dicentra exima

Dictamnus albus

Doronicum orientale

Echinops ritro (center)

Epimedium × rubrum

Eranthis hyemalis

Erigeron hybrid

Eryngium amethystinum

Eupatorium maculatum

Euphorbia polychroma

Gaillardia × grandiflora

Euphorbia myrsinites

Galanthus nivalis

Galium odoratum

Geranium

Gypsophila paniculata

Helenium autumnale

Heliopsis helianthoides

Hemerocallis sp.

Heuchera sanguinea

Hosta sp. (right), *Alchemilla* (left)

Hyacinthus orientalis

Iberis sempervirens

Iris reticulata

Iris sibirica (background)

Kniphofia uvaria

Lamium maculatum

Leucanthemum × superbum

Leucojum aestivum

Liatris spicata

Lilium

Limonium latifolium

Linum perenne (right foreground)

Lupinus hybrid (center background)

Lysimachia clethroides

Lythrum virgatum (rosy-purple)

Monarda didyma

Muscari armeniacum

Narcissus

Nepeta × faassenii

Ornithogalum umbellatum

Paeonia lactiflora

Papaver orientale

Phlox paniculata

Phlox subulata

Physostegia virginiana

Polemonium caeruleum

Primula × polyantha

Pulsatilla vulgaris

Puschkinia scilloides

Rudbeckia fulgida

Salvia × superba

Saponaria ocymoides

Scabiosa caucasica

Scilla sibirica

Stachys byzantina

Stokesia laevis

Tanacetum coccineum

Thermopsis caroliniana

Trollius europaeus

Tulipa

Veronica spicata

Veronica austriaca (center)

Vinca minor

Waldsteinia ternata

Yucca filamentosa

SUMMER PERENNIALS

Achillea filipendulina

Fernleaf Yarrow
Asteraceae (Sunflower Family)
○   Background, Summer

Flowers: The yarrows produce a large flat corymb flower head that makes a bold statement in the background of the garden. Each flower head is 4 to 5 inches wide and bright mustard yellow.

Leaves: The leaves of the fernleaf yarrow are finely divided, fern-like, and alternate. These pubescent leaves are gray-green and range in size from 8 to 10 inches in the basal rosette to 3 to 4 inches near the top of the stem. When rubbed, the leaves give off a pungent aromatic odor.

Growing Needs: Fernleaf yarrow does well in a full-sun location. This plant prefers a well-drained soil and is quite drought resistant, disliking wet conditions. Taller ones may need staking.

Harvest it for fresh or dried flowers. Deadhead the brown faded flowers.

Landscape Use:

Height:	3 to 4 feet (4½ feet) in flower
	2 to 3 feet foliage
Habit:	Upright mounded
Spacing:	18 to 24 inches
Texture:	Medium fine — foliage
	Bold — flower

Plant fernleaf yarrow in a sunny background of the border or as a showy center of an island bed. Grow them as a fool-proof fresh or dried cut flower.

Design Ideas: Fernleaf yarrow looks especially nice with perennial blue salvia, both 'May Night' and 'East Friesland'. Gooseneck loosestrife, *Lysimachia clethroides,* and Maltese cross, *Lychnis chalcedonica,* are a showy blend of flower forms and colors with fernleaf yarrow.

Cultivars:

Taller ones —

'Coronation Gold' is 3 feet with attractive 3- to 4-inch wide flowers and a good habit that does not require staking.

'Gold Plate' is large flowered (5 to 6 inches wide) and tall (4 to 4 ½ feet). Striking!

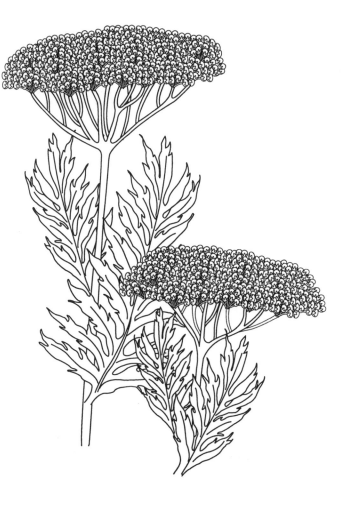

'Parker's Variety' is also tall (4 to 5 feet) with medium-size flowers (3 to 4 inches wide). It may need staking.

Shorter ones —

'Altgold' (Old Gold) is 2 to 3 feet tall and may reflower later in summer.

'Neugold' (New Gold) is 2 feet tall and a profuse flowering plant.

Problems: None serious

Propagation: Seed and division.

Zones: 3-8

Related Species:

Achillea × 'Moonshine' has lemon yellow corymbs, fern-like silvery leaves, a more compact (18 to 24 inches) mounded habit, and earlier flowering (7 to 10 days). It has a moderate secondary flowering. If I could only grow one yarrow, this would be my choice. Nice plant!

A. × 'Anthea' is a newer cultivar with lighter yellow flowers than Moonshine. The leaves are also silvery. May reflower on the secondary stems, especially if harvested or deadheaded, 18 to 24 inches.

Achillea millefolium cultivars, Rose Yarrow, has pink to rose-red, flat-topped flowers of 1 to 3 inch width. Rose yarrow is 1 to 2 feet tall with very finely divided foliage. This plant spreads rather quickly and has a tendency to sprawl. (The straight species is a white flowering native roadside plant.) The flowers can be harvested as a fresh cut flower and can be dried without fading, only if kept cool (near an air conditioner or in a cooler).

'Cerise Queen' has rose pink flowers, fading pink.

'Crimson Beauty' has deep rose pink flowers, also fading with age.

'Fire King' has deeper rose red flowers with a grayish cast to the foliage.

'Lilac Beauty' has lilac flowers that fade to a pastel pink.

'Red Beauty' has crimson red flowers.

'Rosea' is an old standby with pink flowers.

'Rubra' has deeper pink flowers.

'White Beauty' has white flowers.

There are many nice new hybrids with a range of colors, such as peach, red, lavender, and bicolors, and less invasive character.

'Appleblossom' has lilac-pink flowers and is taller at 24 to 30 inches.

'Heidi' has bright pinkish red flowers, 30 inches.

'Orange Queen' has peach to light orange-colored flowers, 28 to 30 inches.

'Paprika' has nice red flowers with yellow centers that fade to pink, 3 feet.

'Peach Blossom' has peach to salmon flowers, aging to pastel peach, 24 inches.

'Royal Tapestry' has purple flowers with white centers, 2 feet.

'Snow Taler' is a long flowering pure white, 28 inches.

'Summer Pastels' has a range of colors from lavender, pink, red, white, tan, yellow, and peach colors. A 1990 All America Selection.

Achillea ptarmica

Sneezewort Yarrow, White Tansy
Asteraceae (Sunflower Family)
○, ◑ Midborder, Summer

Flowers: Sneezewort has a very different-looking flower head compared to the other yarrows. It produces small white flower heads (½ inch across) borne in a loose, rounded corymb. The double forms appear somewhat ball-shaped and resemble a large type of baby's breath. The flowers are excellent cut flowers, both fresh and dry.

Leaves: The alternate leaves of sneezewort are 1 to 4 inches long. Each glabrous green leaf is sessile (no petiole) and linear to lanceolate with a finely toothed margin.

Growing Needs: Like the other yarrows, it performs best in sunny, well-drained locations. Sneezewort is also drought tolerant.

Landscape Use:

Height:	1 to 2 feet
Habit:	Upright mounded to sprawling
Spacing:	18 inches
Texture:	Medium fine

Choose the cultivars for nicer habit and double flowers to place in the sunny foreground to midborder. This is one of my favorite cut flowers to dry; it is a reliable baby's breath-looking substitute.

Design Ideas: It combines well with the other yarrows and red hot pokers with *Coreopsis* 'Moonbeam' as an edging to create a garden with hot color combinations for a hot dry site. Place it with perennial flax for a fine-textured spot of white and sky blue. I plant white tansy with midseason tulips because it does a great job of masking the faded tulip foliage in late May.

Cultivars:

'Angel's Breath' is a double white.

'Ballerina' has double white flowers, is compact (10 to 12 inches), and mounded.

'Perry's White' has double white and needs support.

'The Pearl' has a double white flower, is 2 feet, and more mounded.

'Unschuld' is a double white.

Problems: None serious.

Propagation: Seed or division or summer tip cuttings.

Zones: 3-9

Anchusa azurea

Summer Forget-Me-Not, Italian Bugloss
Boraginaceae (Borage Family)
○, ◑   Background, Summer

Flowers: The ¾-inch flowers of summer forget-me-not are blue (to purple), funnel-shaped, and borne in loose cymes. The buds, which are reddish in color, have a bristly pubescense that continues down the stem. This plant provides a nice summer "true blue" flower color.

Leaves: The leaves of the summer forget-me-not are 6 to 10 inches long and oblong to lanceolate. The surface of the leaf is rough and hairy. They may yellow after flowering so it is best to plant a facer plant in front of summer forget-me-not. Leaves are both basal and alternate on the stem.

Growing Needs: Italian bugloss is a full-sun plant and will tolerate light shade. It prefers a well-drained soil. Either stake the taller plants or use shorter cultivars. Italian bugloss will readily self sow.

Landscape Use:

Height:	(1 foot) 2 to 5 feet
Habit:	Upright
Spacing:	18 to 24 inches
Texture:	Medium bold

Italian Bugloss is a midborder to background plant depending on the cultivar selected. Use the short cultivars as showy foreground specimen plants. Mask yellowing leaves with a facer plant, such as Shasta daisies or daylilies.

Design Ideas: This plant looks especially nice with any pink-flowered plant, such as painted daisy or red valerian (*Centranthus*). Combine Italian bugloss with the silver artemisias and yellow yarrows with catmint or lavender in front.

Cultivars:

'Dropmore' is a tall (4 to 5 feet), older selection that always needs staking.

'Little John' is more compact (12 to 18 inches) with dark blue flowers.

'Loddon Royalist' is 3 feet tall with purplish blue flowers.

'Royal Blue' has intense blue flowers and is 3 feet tall.

Problems: None serious.

Propagation: Self sows, cuttings, and division.

Zones: 3-8

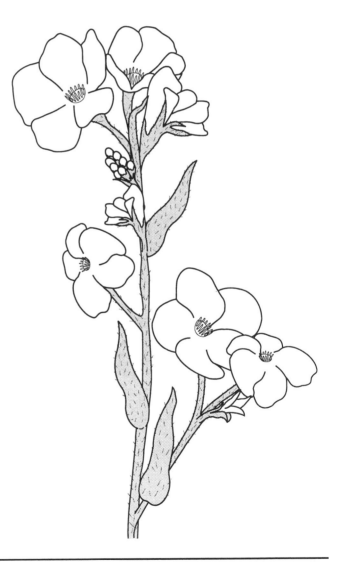

Astilbe × arendsii

Astilbe, False Spirea
Saxifragaceae (Saxifrage Family)
◐, ○   Foreground to midborder, Early summer

Flowers: This showy airy plume is a must for the garden. The flowers are borne in fluffy dense panicles. The 6- to 10-inch upright or arching plumes range in colors from white, pink, peach, red, and lavender.

Leaves: Astilbe leaves are in threes, two to three ternately compound, and alternate. The 1- to 1½-inch leaflets are ovate, toothed, and range from glossy to dull. They are usually green in color but some red varieties have a red to bronze tint to the foliage, especially the newly emerging leaves.

Growing Needs: Astilbe enjoys a moist yet well-drained soil. They can grow in part shade to full sun, although, the leaves may "burn" in full sun if not provided with enough moisture. I have seen drip irrigation used for sunny massed plantings in a city park.

Landscape Use:

Height:	24 to 36 inches
Habit:	Upright mounded
Spacing:	12 to 18 inches
Texture:	Medium fine
	Fine in flower

Astilbe is a partially shaded foreground to midborder plant and makes a remarkable show en masse. It also makes a magnificent specimen plant.

Design Ideas: Group white astilbe with variegated Solomon's seal, Japanese painted fern and the mottled leaves of Bethlehem sage. Combine red astilbe with purple Siberian iris.

Cultivars: Numerous! !

Problems: If under stress, it may get white fly, spider mites, adult Japanese beetle or powdery mildew.

Propagation: Division or seed.

Zones: 4-9

Related Species:

Astilbe chinensis 'Pumila' is a compact 12 to 15 inches tall. This mid-summer one is purplish-pink and can withstand drier conditions.

A. c. var. *taquetii* 'Superba' is the latest astilbe to flower (pink) and is more heat and drought tolerant, 3 to 4 feet.

A. simplicifolia 'Sprite' has pink flowers, is 12 to 18 inches tall, and has finely dissected foliage. 1994 Perennial Plant of the Year

Campanula carpatica

Carpathian Harebell or Bellflower
Campanulaceae (Bellflower Family)
○, ◑ Edging, Summer

Flowers: Delightful 1½-inch, blue or white, cup-shaped flowers arise from graceful slender stalks. The solitary flowers of this campanula are borne over a long period of mid to late summer.

Leaves: The leaves of Carpathian harebell are light green and long petioled. Each leaf is 1 inch long and ovate to triangular in shape with toothed, wavy margins. Both basal and alternate (on the stem) leaves are present.

Growing Needs: This easy-to-grow plant thrives in sunny locations to partly shaded areas. It needs a well-drained soil that is mulched to maintain moisture.

Landscape Use:

Height:	6 to 12 inches
Habit:	Mounded
Spacing:	12 to 18 inches
Texture:	Medium fine

Carpathian harebell makes a good sunny to partially shaded edging plant or as a limited ground cover.

Design Ideas: Combine this bellflower with fall anemone for a contrast in leaf and an overlap in flowering time. Group this delicate flower in partial shade with the boldness of *Heuchera* 'Palace Purple'. It also combines well with *Silene dioica* 'Robin's White Breast' and moonbeam coreopsis as an edging and with *Veronica* 'Goodness Grows' and Siberian iris foliage behind it.

Cultivars:

var. *turbinata* 'Alba' has white flowers and compact growth.

'Blue Clips' has large blue flowers and is 6 to 8 inches tall.

'China Doll' has lavender blue flowers, 8 inches.

'Wedgewood Blue' has light blue flowers and grows 6 inches tall.

'White Clips' has large white flowers, 8 inches.

Problems: None serious.

Propagation: Division and seed

Zones: 3-8

Related Species — Low Growing:

C. portenschlagiana, 'Dalmatian Bellflowers', is a 6 to 8 inch edging plant with light lavender-blue flowers.

C. poscharskyana, 'Serbian Bellflower', is an aggressive 6 to 8 inches edging or ground cover for trailing over walls. The starry flowers are blue or white.

C. × *'Birch Hybrid'* is long flowering and lavender, 6 inches.

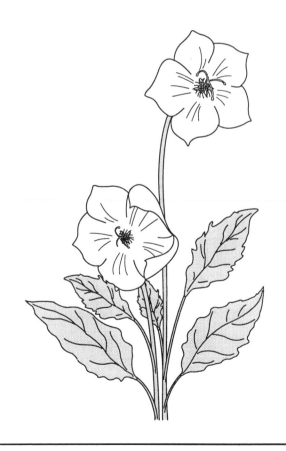

Campanula glomerata

Clustered Bellflower
Campanulaceae (Bellflower Family)
○, ◑ Foreground, Summer

Flowers: Clustered bellflower produces tight clusters of 1-inch, star-shaped purple, blue, or white flowers. Each flower has a deep bell shape. From a distance, these clusters look almost ball-shaped.

Leaves: The alternate leaves of clustered bellflower are 3 to 5 inches, ovate to lanceolate with a finely serrate (serrulate) margin. The medium green leaves have short hispid pubescence.

Growing Needs: Clustered bellflower grows well in a sunny to partly shaded site. A moist well-drained soil is preferred. Clustered bellflower may self sow in the garden.

Landscape Use:

Height:	12 to 24 inches
Habit:	Upright mounded
Spacing:	18 inches
Texture:	Medium

Clustered bellflower looks best in the sunny foreground or midborder of the perennial garden.

Design Ideas: Combine clustered bellflower with Siberian iris and leopard's-bane. Silver king artemisia makes an excellent background plant for the clustered bellflower. Another cheerful grouping for this plant is with The Pearl yarrow and Zagreb threadleaf coreopsis.

Cultivars/Varieties:

'Dahurica' is a true purple and 12 inches.

'Joan Elliot' has dark bluish-purple flowers and is 18 to 20 inches tall.

'Purple Pixie' has small deep purple flowers.

'Superba' has large bluish-purple flowers and is 30 inches tall.

White ones:

'Alba' — 18 inches

'Caroline' — bicolor of violet and white, 20 inches

'Crown of Snow' — 18 to 20 inches

'Schneehaschen' — large flowers, 6 to 8 inches

'Schneekissen' — compact habit

Problems: None serious.

Propagation: Division and seed

Zones: 3-8

Related Species

C. punctata 'Cherry Bells' — large dark rose-pink pendulous bells, ovate leaves, 12 to 18 inches.

C. takesimana — large with white to off white pendulous bells with unique maroon spots, quite assertive, 18 to 24 inches.

Campanula persicifolia

Peach-leaved Bellflower
Campanulaceae (Bellflower Family)
○   Foreground, Midborder, Summer

Flowers: The peach-leaved bellflower has shallow, cup- or bell-shaped flowers borne on an elongated raceme. Each flower is 1 to 1½ inches across and blue or white in color. Remove spent flowers for longer flowering and sparse reflowering.

Leaves: Peach-leaved bellflowers produce a rosette of thin, linear to lanceolate leaves, resembling those of a peach tree. Each 4- to 6-inch leaf is glossy green in color with a noticeable mid-vein. The stem leaves are produced alternately along the stem and become progressively smaller near the top

Growing Needs: Grow peach-leaved bellflower in a sunny to lightly shaded area. A well-drained soil is best for this plant. Remove faded flowers to encourage later flowering.

Landscape Use:

Height:	2 to 3 feet
Habit:	Upright or two tiered
Spacing:	18 inches
Texture:	Medium

Grow peach-leaved bellflower massed in the sunny foreground to midborder. It is a little tricky to site. The 8- to 10-inch mound of foliage would place it in the edging to foreground, but the 2- to 3-foot flowers warrant its placement in the midborder (or as a tall foreground plant). It is also a graceful cut flower. Reliable and well worth growing!

Design Ideas: Plant with *Penstemon* 'Husker Red' and pinks. Combine peach-leaved bellflower with pink soapwort (*Saponaria ocymoides*), Elijah blue fescue, and bearded iris.

Cultivars:

'Alba' has a white flower, 2 feet.
'Grandiflora Alba' has white flowers and grows 2½ feet tall.
'Telham Beauty' has larger (2 inches across) light blue flowers, 3 to 4 feet tall. Excellent!

Doubles/Semi-Doubles:

'Boule de Neige' — large white double.

'Coerulea Coronata' — purplish-blue semi-double.

Problems: None serious.

Propagation: Division and seed.

Zones: 3-7

Centranthus ruber

Red Valerian, Jupiter's Beard, Keys of Heaven
Valerianaceae (Valerian Family)
◯, ◑ Midborder, Summer

Flowers: The ½-inch long tubular flowers are clustered into large heads (2 to 3 inches wide) on the stem. These fragrant flowers can be reddish-pink, pink, or white. The terminal cluster may flower as early as late spring, Other flower clusters are borne in pairs at each leaf axil, which adds up to a long flowering time. If faded flowers are trimmed back, flowering will continue until fall.

Leaves: The lovely green to blue-green leaves are glabrous, ovate, opposite, and 2 to 4 inches long. They are quite attractive throughout the growing season.

Growing Needs: Grow red valerian in a full sun to a partly shaded and slightly alkaline location. This plant can tolerate drought and poor soils, producing more compact plants. This plant readily self sows; trim back to keep in check and to promote more flowering.

Landscape Use:

Height:	2 to 3 feet
Habit:	Mounded
	Mounded sprawling
Spacing:	18 to 24 inches
Texture:	Medium

Plant red valerian in the sunny foreground to midborder. It will also grow well in rock gardens or even draping over stone walls.

Design Ideas: Group this pink to deep rosy-pink midborder plant with soapwort for edging, perennial flax in the midborder, and foxglove for background. *Centranthus* looks nice with lamb's ear and moonbeam coreopsis in front and Russian sage and clematis (on a shrub or trellis) behind it.

Cultivars:
'Alba' — white
'Atrococcineus' — deep red
'Roseus' — rose

Problems: None serious.

Propagation: Seed, cuttings. Transplanting of very young self-sown seedlings is a sure-fire method; division of the fleshy roots is nearly impossible.

Zones: 4-8

Coreopsis grandiflora (and lanceolata)

Coreopsis, Tickseed, Lance Coreopsis
Asteraceae (Sunflower Family)
○, ◐ Foreground, Midborder, Summer

Flowers: These 1½- to 2-inch, yellow flower heads arise on long stems (peduncles). The eight, daisy-like rays are rounded and pinked or jagged at the ends. Before opening, the golden buds are glossy and subtended by a starry collar of sepals.

Leaves: The leaves of coreopsis are opposite, glabrous, and green. Each 2- to 6-inch leaf is oblong to lanceolate and often polymorphic with small lateral lobes (1 to 4). The variety of leaf shapes (like the sassafras tree) makes this an easy identification feature. *C. lanceolata* has larger leaves that are rarely lobed.

Growing Needs: This plant will thrive in a sunny, well-drained location. Remove faded flowers to promote reflowering and to keep self sowing to a minimum.

Landscape Use:

Height:	1 to 3 feet
Habit:	Mounded to sprawling
Spacing:	12 inches
Texture:	Medium

Coreopsis fits nicely into the sunny foreground or midborder. Tickseed works well in a cottage garden situation, as well as a cutflower garden or an informal naturalized area.

Design Ideas: Tickseed makes a nice splash of color in the garden when planted with clustered bellflower, giant allium and delphinium. It also combines well with purple coneflower and garden phlox.

C. grandiflora Cultivars:

'Domino' is 15 inches and has single yellow flowers with a dark reddish-brown center.

'Early Sunrise' is compact, 18 to 20 inches, with double flowers. This 1989 All-American Selection can be grown true to type from seed.

'Sunray' has large (3 to 4 inches) double yellow flowers and grows 2 feet tall.

C. lanceolata Cultivars:

'Baby Sun' is 12 to 18 inches with single yellow flowers with dark red centers.

'Goldfink' is a compact 8 to 10 inches with nice single yellow flowers.

Problems: None serious.

Propagation: Division, seed, (self sows)

Zones: 4-9

Coreopsis verticillata

Threadleaf Coreopsis
Asteraceae (Sunflower Family)
○, ◑   Foreground, Midborder, Summer

Flowers: The bright golden to lemon yellow flowers are daisy-like but with thinner, elongated, pointed petals (rays). The 1- to 2-inch wide flowers are borne in loose corymbs and bloom profusely over a long period of time.

Leaves: This beautiful finely textured foliage is opposite and palmately dissected into threes with each section again divided into very linear (thread-like) sections. Each 2- to 3-inch leaf is medium green and attractive all through the growing season.

Growing Needs: Site threadleaf coreopsis in a sunny, well-drained location. This very attractive plant is drought tolerant and not as likely to self sow as *Coreopsis grandiflora*. Remove spent flowers to promote reflowering.

Landscape Use:

Height:	2 feet
Habit:	Mounded
Spacing:	18 inches
Texture:	Fine

Threadleaf coreopsis is an excellent softening plant for the garden. Plant it near bold or coarse textures, such as globe centaurea or oriental poppies, to contrast and soften them. Use it as a sunny foreground to midborder plant. Moonbeam can be an edging plant and a limited ground cover.

Design Ideas: The bright golden yellow ones look great with the orange-red of Maltese cross (*Lychnis chalcedonica*) and red hot pokers. For a cheerful combination, plant the pink showy primrose and moonbeam coreopsis together. All threadleaf coreopsis contrast well with both annual and perennial blue salvia.

Cultivars/Related Species:

'Grandiflora' ('Golden Shower') is a cheerful golden yellow flowering, 2- to 3-feet tall plant.

'Moonbeam' is a mounded, 18 inches tall with lemon-yellow flowers. 1992 Perennial Plant of the Year

'Zagreb' is a compact 12 to 18 inches with golden yellow flowers.

Coreopsis rosea 'Nana' is a pink flowering coreopsis that is 12 to 18 inches tall and quite a vigorous spreader.

Problems: None serious.

Propagation: Division, seed (may self sow).

Zones: 3-8

Delphinium elatum

Delphinium
Ranunculaceae (Buttercup Family)
○, ◑   Midborder, Background, Early to
mid-summer

Flowers: Delphinium produces quite showy 6-
to 12-inch racemes, generally one main raceme per
plant with some secondary side racemes. Each 2- to
3-inch floret is composed of five petaloid sepals
with a spurred one as well as a central "bee" of
smaller darker petals. Delphinium is most commonly blue but can also be white, violet, pink, red,
bicolors, and single and double flowered.

Leaves: The medium green leaves emerge in
early spring and are soft and pubescent. These attractive leaves can be large (4 to 8 inches) and are
palmate with five to seven deep lobes. The basal foliage persists throughout the summer and fall. The
stem leaves are alternately arranged.

Growing Needs: Delphiniums are the most
floriferous in the sun but will tolerate light shade.
They prefer a well-drained, humus-rich, slightly alkaline, fertile soil. Deeply prepare the soil, add organic matter, and mulch for the best results. Taller
flowers may need staking. In spring, select four to
five of the thickest, strongest stems on well-
established plants and pinch out the weaker stems.
This practice will produce high-quality flowers and
plants. After flowering, cut back any faded flowers
to promote a second flush of flowers.

Landscape Use:

Height:	2 to 6 feet
Habit:	Upright to upright mounded
Spacing:	2 to 3 feet
Texture:	Medium to medium bold

Delphinium makes an attractive specimen in the
sunny midborder to background and is excellent
in the cutting garden. This plant is also lovely
when massed. Although short-lived in our climate,
its beauty makes it well worth the effort.

Design Ideas: Plant delphiniums with Shasta
daisy and painted daisy or with garden phlox and
coral bells. Imagine a garden arbor planted on either side with masses of delphinium (background), bee balm (midborder), and red valerian
(foreground). The bee balm will cover any empty
delphinium spots in the coming years.

Cultivars/Hybrids:

'Blackmore and Langdon' hybrids are hardy, tall,
   6-foot pillars of rich purple, blue, lavender,
   white, and other shades. Very imposing! Need
   staking.

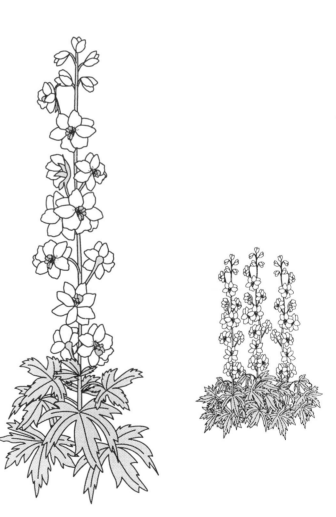

'Connecticut Yankee' have graceful single florets on well-branched racemes, which flower profusely in a wide array of colors, 2½ feet tall.

Dwarf Pacific Hybrids (shorter—2 feet, not long lived, often treated as an annual).

'Blue Fountains' is 2 feet tall in mixed blues.

'Blue Heaven' has sky blue flowers and is 2 feet tall.

Giant Pacific Hybrids are 4 to 5 feet tall in a wide range of colors of mostly double flowers and are easily produced by seed.

'Blue Bird' has medium blue flowers with a white bee.

'Blue Jay' has medium blue flowers with a darker bee.

'Summer Skies' has shades of light to dark blue with a white bee, 5 to 6 feet.

The Round Table Series includes:

'Astolat' has lavender rose shades with a contrasting eye, 5 to 6 feet tall.

'Black Knight' has deep purple flowers with a black bee, 5 to 6 feet tall.

'Galahad' has white flowers, 5 to 6 feet tall.

'Guinevere' has lavender pink flowers with a white bee and is 4 feet tall.

'King Arthur' has purple flowers with a white bee, 4 to 5 feet tall.

'Magic Fountains' are 2 to 3 feet tall with a wide range of colors, often with contrasting eyes. Reflowering is common if faded flowers are removed.

Problems: Cyclamen mites, stalk borer, and slugs. Foliar bacterial leaf spot, root rot, crown rot, bacterial bud rot, botrytis, and powdery mildew. Site well and provide good air circulation.

Propagation: Seed (will usually flower first year), cuttings in spring.

Zones: 3-7

Related Species:

D. × *belladonna*, Belladonna Delphinium, are shorter at 3 to 4 feet, have multiple flowering stalks, and more finely dissected foliage.

'Bellamosum' has dark blue flowers.

'Blue Shadow' has deep blue flowers with a purplish stem.

'Casablanca' has pure white flowers.

'Cliveden Beauty' has pale blue flowers.

D. *grandiflora* 'Blue Butterfly' is a compact 12 inches with showy bright blue flowers.

Dictamnus albus

Gas Plant
Rutaceae (Rue Family)
○ Midborder, Early summer

Flowers: The 1-inch flowers are white (or pink) in a showy terminal raceme. Each flower is composed with five showy irregular petals and prominent curving stamens. (Linnaeus discovered that the volatile oil near the flowers and main stem would ignite, hence the common name.)

Leaves: The pinnately compound leaves of the gas plant are similar to those of an ash tree (*Fraxinus*) with as many as 11 leaflets and are alternate on the stem. Each glossy dark green leaflet is ovate and 2 to 3 inches long. When rubbed, the leaves have a lemon-like fragrance.

Growing Needs: After planting in full sun in a well-drained, moderately fertile soil, there is little maintenance of the plant. Gas plant does not tolerate transplanting so plant it in its permanent spot in the garden. The seeds of this plant are poisonous. In hot weather, contact with the foliage may cause dermatitis.

Landscape Use:

Height:	2 to 3 feet
Habit:	Upright mounded
Spacing:	2 to 3 feet
Texture:	Medium bold (flower)
	Medium (foliage)

Gas plant makes an excellent specimen in the garden. It can be planted in the sunny midborder or background. The clump will improve with age.

Design Ideas: Plant a white one with a fernleaf peony and yellow corydalis. Or, combine a pink gas plant with bearded irises and bellflowers. Gas plant combines well with delphinium and Siberian catmint as background plants, pink or near white daylilies in the midborder with it, and hardy geraniums ('Striatum') and tricolor ornamental sweet potato as edging.

Cultivars:

'Purpureus' (var. purpureus or 'Rubra') has striking pink flowers with dark red veins.

Problems: None serious.

Propagation: Seed, purchase small containerized plants.

Zones: 2-9

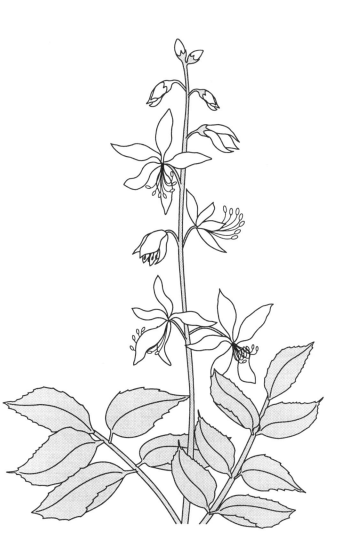

Erigeron hybrids

(E. speciosus, glaucus)

Fleabane
Asteraceae (Sunflower Family)
○   Foreground, Mid-summer

Flowers: These 1- to 2-inch flower heads are daisy-like (aster-like) with yellow centers. They are solitary or borne in a corymb and range in color from pinks, blues, lavenders, and whites. This is the summer aster!

Leaves: Fleabane produces green to gray-green basal leaves that are 3 to 4 inches long as well as sessile stem leaves of progressively smaller size. The leaf shape is lanceolate; the leaf surface is glabrous with some pubescence along the margins of the leaves.

Growing Needs: *Erigeron* grows well in full sun, well-drained sites with low to moderate fertility. This plant is drought resistant. Remove faded flowers to extend the season with a sparse reflowering.

Landscape Use:

Height:	1 to 2 feet
Habit:	Mounded
Spacing:	12 to 18 inches
Texture:	Medium

Fleabane looks best when placed in the sunny foreground or in front of a shrub border. Use it in a cut flower, cottage, or rock garden.

Design Ideas: Try pink fleabane with *Geranium* 'Johnson's Blue' and *Ajuga* 'Burgundy Glow'. Also, combine it with mountain bluet (*Centaurea montana*) and coral bells. Combine fleabane with other drought-resistant plants, such as yarrows, purple coneflowers, and liatris.

Cultivars:

'Azure Blue' has large lavender-blue flowers, 2½ feet tall.

'Darkest of All' has deep bluish-purple flowers, 18 to 24 inches tall.

'Elstead Pink' has pale pink flowers.

'New Summer Snow' has white flowers, blushed with pink, 2 feet tall.

'Prosperity' has lilac blue flowers, 18 inches tall.

'Rosa Triumph' has semi-double, deep pink flowers, 2 feet tall.

Problems: None serious. If poorly sited, powdery mildew, leaf spots, rusts.

Propagation: Seed, fall division, spring vegetative stem cuttings.

Zones: 3-8

Gaillardia × grandiflora

Blanket Flower
Asteraceae (Sunflower Family)
○   Foreground, Midborder, Summer

Flowers: These showy, 2- to 4-inch flower heads are solitary and vibrant in yellows, reds, or banded combinations of the two. The colorful ray flowers are three-toothed at the tips. Rounded puffs of faded flower centers (disk flowers) persist on the plant.

Leaves: The light green to gray-green leaves are pubescent and regularly lobed with a light midvein. The 8- to 10-inches leaves are borne in a basal rosette; the stem leaves are petioled, lobed and alternate on the stem although the upper leaves are sessile and entire (that is, no lobes).

Growing Needs: This easy-to-grow plant thrives in full sun and well-drained, low to moderately fertile soil with high organic matter content. Site for good air circulation to prevent mildew. Taller ones may need peastaking. Remove the earliest faded flowers to promote more flowering.

Landscape Use:

Height:	2 to 3 feet
Habit:	Mounded to sprawling
Spacing:	12 inches
Texture:	Medium bold to medium

Use blanket flower massed in the sunny foreground to midborder. This is a popular plant because of its long season of flowering. It makes a good addition to the cut flower garden.

Design Ideas: This plant looks festive with threadleaf coreopsis and red hot poker. Baby's breath softens the coarseness of blanket flower. The shorter edging ones look great in front of veronica or perennial blue salvia.

Cultivars:

'Baby Cole' has red flowers that are tipped in yellow, 8 inches tall.

'Burgundy' has large wine red flowers and grows 24 to 30 inches tall.

'Dazzler' is bright yellow with dark red centers and 2 to 3 feet tall.

'Goblin' ('Kobold') is a showy 1 foot tall with red flowers tipped in yellow.

'Golden Goblin' ('Goldkobold') is completely golden yellow, 12 inches.

'Monarch Strain' is combinations of yellow and red, 2 to 3 feet tall, seed grown.

Problems: Powdery mildew.

Propagation: Division, cuttings. Seed for Monarch Strain only. (Some may self-sow.)

Zones: 4-9

Gypsophila paniculata

Baby's Breath
Caryophyllaceae (Pink Family)
○, ◑ Midborder, Summer

Flowers: These tiny, ¼-inch, white flowers can be single or double and are borne in loose rounded panicles. Baby's breath makes an excellent softening flower as well as cut flower, both fresh and dried.

Leaves: Each small, 1- to 3-inch, blue-green leaf is waxy and lanceolate. The leaves are opposite and resemble those of carnations.

Growing Needs: This plant grows well in the sunny to partially shaded perennial border. It does best in a deep, moist, well-drained, slightly alkaline soil. If baby's breath is languishing in your garden, check the soil; it probably needs lime.

Landscape Use:

Height:	18 to 36 inches
Habit:	Mounded
Spacing:	2 feet
Texture:	Fine

This plant makes an excellent midborder facer plant; it helps hide unsightly leaves (*Anchusa*), masks disappearing acts of some plants (*Papaver)*, and will also soften the appearance of bolder plants (*Centaurea macrocephala* or *Gaillardia*). Place it in a cut flower garden.

Design Ideas: It is great to mask the fading foliage of poppies or spring bulbs (tulips, daffodils, hyacinths) in summer or to soften perennial bachelor button. Baby's breath is great with garden iris and daylilies.

Cultivars:

'Bristol Fairy' is 2 feet tall with double flowers.

'Perfecta' is a nice large double flower that is 2½ feet tall.

'Pink Fairy' is a soft pink, double flowering 18 inch plant.

'Snow Flake' has mostly double white flowers and grows 3 feet tall.

'Viette's Dwarf' is a compact 14 to 16 inches with pale pink double flowers.

Problems: None serious.

Propagation: Cuttings, seed. (Commercially by grafting.)

Zones: 3-9

Related Species:

Gypsophila repens is a creeping form of baby's breath. It is only 4 to 6 inches tall and will spread 2 to 3 feet. The pink or white ¼-inch flowers are single forms. 'Rosea' is a lovely pink.

'Dorothy Teacher' is also pink.

Hemerocallis sp.

Daylily
Liliaceae (Lily Family)
○ Midborder, Background,
Summer (early, mid, late season)

Flowers: Each trumpet-, star-, or flat-shaped, 2- to 6-inch flower has three inner petals and three outer sepals. Each erect, slender flowering scape produces 5 to 15 flowers. Almost every color is available to grow—choose a variety for a long season of flowering! A tremendous variety of flower types (ruffled, recurved, spider, triangular, flaring, pinched, eyed, contrasting midribs, etc.) are also available. Many types are fragrant; some are repeat flowerers.

Leaves: The leaves arise from a basal rosette. They are long, linear and medium green, arching to 12 to 18 inches. Some daylily cultivars may be evergreen while the rest are deciduous.

Growing Needs: This is a long-lived perennial in the garden. It prefers a full sun location and well-drained soil. Add organic matter to sandy or clayey soils. Light colors may fare better in a lightly shaded location. For the neatest appearance, remove the spent flowers regularly; trim off the flower scapes after flowering.

Landscape Use:

> Height: 1 to 4 feet
> Habit: Vase shaped, two tiered
> Spacing: 24 inches
> Texture: Medium

Plant this showy plant massed or grouped in threes in the sunny midborder to background.

Design Ideas: Use daylilies interplanted with daffodils or Alaska Shasta daisy. It looks great with black-eyed Susan and *Geranium sanguineum* 'Alba'. Combine poppies and daylilies together so the daylilies will cover the unsightly fading summer foliage of the poppies. Self-sowing rose moss makes a nice full sun edging for daylilies.

Cultivars: Numerous! Something for everyone!

The long-flowering cultivars include 'Stella d'Oro' and 'Happy Returns'. 'Stella d'Oro' has a golden yellow flower and is 12 to 15 inches tall. 'Happy Returns' has a lemon yellow flower and is also 12 to 15 inches tall.

Problems: None serious.

Propagation: Division after flowering in July to August. Separate roots gingerly. Roots can be divided down to one section (of roots), if you are a patient gardener, or two to three root sections to fill in faster. Trim foliage back to a 4- to 6-inch fan, plant, and mulch.

Zones: 3-9

Heuchera sanguinea

Coral Bells
Saxifragaceae (Saxifrage Family)
○, ◐ Edging, Foreground, Early Summer

Flowers: The flowers arise on graceful slender stems. Each ¼- to ½-inch, funnel-shaped flower is borne on narrow panicles. The flowers range in color from red to pink to white and are quite attractive to hummingbirds.

Leaves: The basal rosette of leaves produce 2-inch, rounded, lobed leaves in green, purple, or silver. The leaves can be mottled with a darker green or silver. The leaves can be semi-evergreen.

Growing Needs: Plant coral bells in a full sun, well-drained location. The darker leaved hybrids benefit from partial shade. Add organic matter for the best growth. Remove faded flowers for additional flowering. In colder climates with little snow cover, mulch plants well before winter to avoid frost heaving.

Landscape Use:

Height:	18 inches in flower
	6 to 8 inch basal rosette
Habit:	Two tiered
Spacing:	12 to 18 inches
Texture:	Fine to medium fine

Coral bells make an excellent edging to foreground plant. It also makes a nice addition to the cut flower garden.

Design Ideas: I do enjoy my grouping of 'Firesprite' (rosy red) coral bells with fragrant lavender and fine-textured *Geranium sanguineum* 'Alba' along my front sidewalk. The textures and flower colors/types blend well. Try pink coral bells with *Dianthus* 'Spottii', catmint, and columbine. The purple leaved *Heuchera micrantha* 'Palace Purple' looks beautiful with pink soapwort underneath a Prairifire crabapple with its matching burgundy new growth.

Cultivars:

'Splendens' have dark red flowers.
'Variegata' has variegated leaves.

Hybrids:

H. × brizoides (*H. sanguinea* is one parent).
'Bressingham Hybrid' has a range of red, pink, coral, or white flowers and grows 22 inches tall. 'B. Blaze' is a red one.
'Chatterbox' is pink, long-flowered, and 18 inches tall.

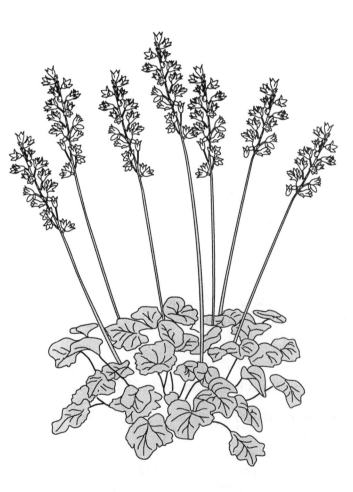

'Firebird' has dark red flowers, 18 inches tall.

'Frosty' is a red flowering, silver variegated, 20 inches.

'June Bride' is 15 inches and white.

'Matin Bells' has bright coral red flowers and is long flowering, and 18 inches tall.

'Plue de Feu' ('Rain of Fire') has bright cherry red flowers.

'Red Spangles' is a very dark red with lightly mottled leaves.

'Scintillation' has bright pink flowers that are tipped with red, 2 feet tall.

'Snowstorm' has noticeable white flecking and cerise flowers, 18 inches tall.

Problems: None serious.

Propagation: Division and seed.

Zones: 3-8

Related Species:

H. americana is grown more for its foliage effect than for its flowers.

'Dale's Strain' has silvery, blue-gray foliage and is a compact 12 to 14 inches.

'Garnet' is a deep garnet-red leaved coral bell.

'Pewter Veil' has dark purple leaves mottled with silver and reddish pink.

'Ruby Veil' has large gray leaves with red and purple mottling and dark purple venation.

H. micrantha 'Bressingham Bronze' has bronze leaves with crimped edges and is 18 inches tall.

'Chocolate Ruffles' has large ruffled foliage with chocolate brown on top and dark burgundy underneath, 30 inches tall.

'Montrose Rose' has rounded leaves of dark red and some silver mottling, 18 inches tall.

'Palace Passion' is similar to Palace Purple but with showy pink flowers.

'Palace Purple' is the break through plant for dark purple leaves and a 1991 Perennial Plant of the Year. It has dark purple, bronze, or greenish purple maple-like leaves with ivory flowers. Partial shade is beneficial.

'Pewter Moon' has dark burgundy leaves with pewter gray mottling, 1 foot tall.

'Plum Pudding' is plum purple leaved with a nearly glossy finish. Flowers are creamy white, 22 inches tall.

X Heucherella alba (H. × brizoides X Tiarella wherryi) is a clump-forming hybrid with coral bell-like flowers above foam flower-like foliage.

'Bridgette Bloom' pink flowers, 1 to 2 feet tall.

'Pink Frost' pink flowers, 12 to 18 inches tall.

'Rosalie' is light pink with a dark mottled leaf, 15 inches tall.

'Crimson clouds' has a small leaf inset in the center of the crimson mottled leaf and has pink flowers, 16 to 18 inches tall.

X Heucherella tiarelloides (H. × brizoides X Tiarella cordifolia) is a pink flowering ground cover, 18 inches tall.

Iris sibirica

Siberian Iris
Iridaceae (Iris Family)
○, ◑   Midborder, Early Summer

Flowers: The flowers come in blue, purple, yellow, white, and bicolors. The 2- to 3-inch flower is divided into six segments, three standards and three falls. The petals are veiny and strappier than garden iris and more three-dimensional than Japanese iris.

Leaves: The basal leaves are linear, narrow, and 2 to 3 feet long. The blue-green leaves remain attractive all growing season, persist through winter (brown), and need to be trimmed back in early spring for more vigorous growth.

Growing Needs: Grow Siberian iris in a sunny or partially shaded, well-drained location. A moist, fertile, humus-rich, slightly acid soil is beneficial, but Siberian iris will also tolerate dry soils. Divide every 7 to 10 years to rejuvenate.

Landscape Use:

Height:	2½ to 3 feet
Habit:	Upright, vase shaped
Spacing:	18 inches
Texture:	Medium to medium fine

Grow Siberian iris massed in the midborder or use it as an accent.

Design Ideas: Purple Siberian iris combines well with lavender chives. The foliage of Siberian iris adds a nice vertical accent to lady's mantle, hardy geraniums, and asters throughout the summer.

Cultivars: Numerous!

'Ann Dasch' is a showy purplish-blue mottled with a paler whitish-blue.

'Butter and Sugar' is a yellow and white bicolor.

'Caesar's Brother' is a tall, vigorous dark blue, 3 to 3½ feet tall.

'Pirate Prince' is my favorite dark purple iris (by Steve Varner) with standards held at a 45-degree angle.

'White Swirl' has a white flower.

Problems: None serious. (No iris borer)

Propagation: Division.

Zones: 3-9

Related Species:

Iris ensata, Japanese Iris, is similar to the Siberian iris. The 4- to 6-inch flower is flattened, two-dimensional, and blooms a little later than Siberian Iris. It also prefers moist, acidic, humusy soil and is excellent near water.

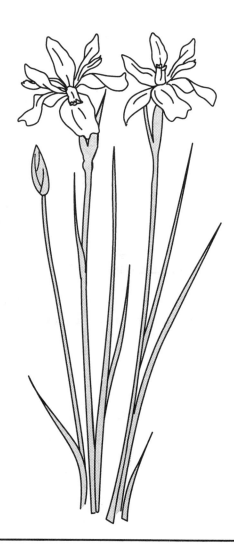

98

Leucanthemum × superbum

(Chyrsanthemum × superbum)

Shasta Daisy
Asteraceae (Sunflower Family)
○, ◐   Foreground, Midborder, Early to
mid-summer

Flowers: These daisy flower heads are white with yellow centers. Each flower head is 2 to 4 inches wide. Some cultivars are double or semi-double. Cut back after flowering for a sparse re-flowering.

Leaves: The attractive, 8- to 12-inch green leaves are glabrous and glossy. The leaf shape is lan-ceolate to spatulate with serrate leaf margins. The leaves of the basal rosette are larger than the stem leaves.

Growing Needs: These plants do well in full sun to partial shade. The soil should be moist, fer-tile, and well drained. Shasta daisy is a heavy feeder — fertilize in spring and after flowering. Mulch them to avoid frost heaving. Division will help this perennial live longer.

Landscape Use:

Height:	1 to 3 feet
Habit:	Mounded
Spacing:	12 to 18 inches
Texture:	Medium

Shasta daisy stands alone or in mass in the sunny foreground to midborder of the perennial garden. It also makes a good addition to the cut flower garden.

Design Ideas: Shasta daisy looks beautiful with astilbe and delphinium in partial shade. It also combines well with threadleaf coreopsis, white gaura, and Siberian iris.

Cultivars:

'Aglaya' has unique frilled double flowers (3 inches across) and is 2 feet tall.

'Alaska' has large, 3-inch, single white flowers and is 2 to 3 feet tall. A classic!

'Becky' is a sturdy 3 to 4 feet tall, long flowering, and a large (4 inch wide) daisy.

'Little Miss Muffet' is shorter, 8 to 12 inches, and semi-double.

'Marconi' has showy, 3- to 5-inch wide, semi-double flowers and reaches 3 feet tall.

'Polaris' is a large flowering daisy (5 to 6 inches wide) and 2 to 3 feet tall.

'Silver Princess' is a nice foreground Shasta daisy, growing 14 inches.

'Snowcap' is long flowering and 12 to 15 inches.

'Summer Snowball' has early, double, 3-inch flow-ers and is 2½ feet tall.

'Thomas Killin' has a tufted center, 2 feet.

Problems: Leaf spot, stem rot, aphids.

Propagation: Division and by seed.

Zones: 5-9

Linum perenne

Perennial Flax
Linaceae (Flax Family)
○, ◑ Foreground, Early to mid-summer (late spring)

Flowers: Perennial flax produces multi-branched panicles covered with sky blue flowers. The 1-inch flowers each last only one day but are borne in profusion. The buds are pendulous but turn upward when fully open.

Leaves: The 1-inch leaves are narrow, linear, and blue green. They are very light and airy in appearance.

Growing Needs: Perennial flax grows best in full sun to light shade. It prefers a moist but well-drained soil. Mulching is suggested and encourages self sowing.

Landscape Use:

Height:	1 to 2 feet
Habit:	Vase shaped
Spacing:	12 inches
Texture:	Fine

Plant perennial flax in the sunny foreground. Its fine texture has a softening effect on coarser plants around it.

Design Ideas: Combine with pink or white creeping phlox and grape hyacinth. Plant perennial flax with *Ajuga* 'Pink Beauty' and hardy white miniature roses. Perennial flax with *Salvia argentea* (Silver Sage) with its 8-inch woolly white leaves is a study in textural contrast and quite pleasing.

Cultivars:

'Alba' has nearly white flowers, 18 inches.
'Diamant' is white and 12 to 14 inches.
'Saphir' is a compact 12 inches.
ssp. *alpinum*, Alpine Flax, has smaller leaves, blue flowers, and is 1 foot tall.

ssp. *lewisii*, Prairie Flax, has larger leaves and is taller at 2 to 3 feet.

Problems: None serious.

Propagation: Seed, cuttings.

Zones: 4-8

Related Species:

L. narbonnense 'Heavenly Blue' has slightly larger and darker blue flowers that are longer lasting on 18-inch stems. Very nice plant!

'Album' is the white form.

'Six Hills' is a sky blue.

Lupinus hybrids

Lupine
Fabaceae (Legume Family)
◑ (○)   Background, Early summer

Flowers: Each showy raceme is 8 to 12 inches long. The individual florets are legume-like or butterfly-like (papilionaceous). The flower is ½ inch in size and has winged and keeled parts. The popular 'Russell Hybrids' and other mixes come in every color!

Leaves: These easily identified leaves are alternate, long petioled, and palmately compound with 9 to 15 leaflets. Each 2- to 6-inch medium green leaflet is a pointed lanceolate shape. The leaves are glabrous on top with pubescence on the underneath side.

Growing Needs: Lupines perform relatively well in partial shade but prefer cooler climates in full sun. The soil should be slightly acidic, light, and evenly moist. Fertilization and mulch would benefit this plant. Site well for the least stress on the plant.

Landscape Use:

Height:	3 to 5 feet
Habit:	Upright mounded
Spacing:	2 feet
Texture:	Medium

Lupines make a nice partially shaded midborder to background planting. One plant or small groupings of three or four of the same color make an excellent show. They are also good cut flowers. During hot and dry summers, lupine foliage can look awful; place plants nearby to mask this potential problem.

Design Ideas: Plant light pink lupines with dark rosy pink painted daisies. Try combining lupines with taller hardy geraniums or with later-flowering fall anemones (to mask late summer lupine foliage).

Cultivars/Hybrids:

Minarette Mix is 15 to 20 inches tall with a range of colors.

Popsicle Mixture is a sturdy 18 to 24 inches tall with a variety of colors.

Russell Hybrids are more heat-tolerant plants with a wide range of flower colors available, 3 to 4 feet.

'Chandelier' — yellow flowers, 30 inches.

'My Castle' — brick red flowers, 30 inches.

'The Governor' — blue/white, 30 inches.

Problems: None serious. Occasionally, rust, powdery mildew.

Propagation: Seed. Dust seeds with nitrifying powder. (Division if plant lives long enough.)

Zones: 4-6

Lychnis chalceodonica

Maltese Cross
Caryophyllaceae (Pink Family)
○, ◑ Midborder, Summer

Flowers: The vibrant scarlet red flowers have five, ½- to 1-inch notched petals. These flowers really catch the eye! The flowers are clustered in dense terminal heads of 3 to 4 inch widths with 10 to 50 flowers!

Leaves: The 2- to 4-inch leaves are opposite and ovate to lanceolate. The leaves clasp the stems, both having hispid pubescence.

Growing Needs: This plant is a striking addition to sunny to partially shaded, well-drained, moderately fertile gardens. Although somewhat short-lived, Maltese cross will easily self sow and perpetuate itself.

Landscape Use:

Height:	2 to 4 feet
Habit:	Upright mounded
Spacing:	12 inches
Texture:	Medium

Maltese cross is a wonderful addition to the old-fashioned or cut flower garden. Place it in the sunny midborder in small groupings.

Design Ideas: Plant with silver-leaved plants, like lamb's ear or artemisia. Yellow and orange plants like yarrow, coreopsis, and globe centaurea combine well with it. Use caution in combining other reds with it; Maltese cross has a orange undertone and may clash with purplish-reds. Bicolor (orangish red and yellow) red hot pokers, perennial blue salvia, and Maltese cross make a striking trio.

Cultivars:

'Alba' has white flowers.
'Flore Plena' is a double scarlet red.
'Grandiflora' is a large scarlet red.

Problems: None serious. Occasionally, leaf spot and rust.

Propagation: Seed and division.

Zones: 3-9

Related Species:

Lychnis coronaria, Rose Campion, has graceful single rounded vibrant purple-red flowers. The lovely basal foliage is silvery and woolly. This airy, two-tiered plant is 1½ to 2 feet tall and does well in the sunny midborder of an old-fashioned garden. Rose campion readily self sows.

'Alba' is a nice white flowering plant.

'Angel's Blush' has a rosy pink blush in the center of the white flower.

L. × arkwrightii 'Vesuvius' has bright orange-red flowers above dark red foliage and is 12 inches tall. Marginally hardy.

L. viscaria, Catchfly, is a showy car-mine red panicle with sticky stems and leaves (hence, the common name), 15 inches tall.

Monarda didyma

Bee Balm, Bergamot, Oswego Tea
Lamiaceae (Mint Family)
○, ◑   Midborder, Summer

Flowers: The 1-inch long, fragrant, tubular flowers are borne in single or double whorls. The flowers are subtended by leafy bracts. Each flower cluster is 2 to 3 inches across and red, pink, white, or lavender in color.

Leaves: The 3- to 5-inch sessile leaves are thin, ovate to lanceolate, slightly toothed, and pubescent. The stems are square and have a minty smell.

Growing Needs: This plant likes part shade and may be vigorous; full sun locations will keep it in check. It prefers a moist, mulched, well-drained soil. Divide this plant often.

Landscape Use:

Height: 2 to 3 feet
Habit: Mounded
Spacing: 2 feet
Texture: Medium

Bee balm makes an excellent accent plant or a good show as a mass planting in the sunny midborder to background of the garden to attract hummingbirds and bees.

Design Ideas: Plant bee balm in an area where it will naturally be kept in check, for example, interplanted with penstemon and perennial fountain grass with lawn surrounding it. Group red bee balm with delphinium, white wild sweet William, and red coral bells (the perfect hummingbird garden). I like the variety of flower shapes of the "firecracker flowers" with daisies or daylilies.

Cultivars: Many are hybrids of *M. didyma* and *M. fistulosa.*

'Adam' has a red flower on 2 to 3 foot stems.

'Aquarius' is a pink and white bicolor.

'Blue Stocking' ('Blaustrumpf') is a nice purple, 2 to 3 feet.

'Blue Wreath' ('Blaukranz') is purplish red, 3 to 4 feet tall, and has good mildew resistance.

'Bowman' is a purple bee balm with mildew resistance and tolerance of dry conditions, 4 feet tall.

'Cambridge Scarlet' is a 3 foot, rose red.

'Croftway Pink' has light pink flowers, is 3 to 4 feet tall, and is very susceptible to powdery mildew

'Gardenview Scarlet' have large red flowers, is 3 to 4 feet tall, and has some powdery mildew resistance.

'Jacob Kline' is large and red with good mildew resistance, 4 feet tall.

'Mahogany' has deep red flowers, 3 feet tall.

'Marshall's Delight' is long-flowering, bright pink, and mildew-resistant.

'Petite Delight' is 15 to 18 inches, lavender-rose with mildew resistance.

'Prairie Night' is a purplish violet, 3 feet tall.

'Purple Crown' is an early flowering purple flower with purple bracts, 3 feet tall.

'Raspberry Wine' has beautiful rosy raspberry-colored flowers and is qute mildew resistant, 4 feet tall.

'Snow Queen' has creamy white flowers with fair mildew resistance, 4 feet tall.

'Twins' is very dark pink, mildew resistant, and 3 feet tall.

'Violet Queen' is lavender blue, 2 to 3 feet tall, and mildew resistant.

Problems: Powdery mildew.

Propagation: Division and seed.

Zones: 4-8

Related Species:

Monarda fistulosa, Wild Bergamot, is the native species that has a lavender-pink, 2- to 3-inch flower.

Nepeta × faassenii

Catmint
Lamiaceae (Mint Family)
○, ◑ Edging, Foreground, All summer

Flowers: Each ½-inch flower is blue-violet and two-lipped. The tubular flowers are borne on leafy interrupted racemes. Trim back after flowering for prolonged additional flowering.

Leaves: The opposite foliage of catmint is a beautiful pubescent gray green. Each 1- to 1½-inch leaf is narrowly ovate with crenate (scalloped) margins. The leaf and flower color blend together so well. Look for square stems.

Growing Needs: This easy-to-grow plant thrives in full sun. The soil should be well drained. Catmint will tolerate partial shade and competition from tree roots. Some of the taller ones tend to flop and can be subtly peastaked (use branchy tree trimmings placed around the perimeter) early in spring to keep the plant mounded.

Landscape Use:

Height:	1 to 2 feet
Habit:	Mounded
Spacing:	12 to 18 inches
Texture:	Medium fine

Use catmint in the foreground or as an edging or ground cover in the sunny border.

Design Ideas: Combine catmint with European columbine and pinks. Its gray-green foliage looks nice with woolly thyme and silver plants, such as snow in summer or beach wormwood (*Artemisia stelleriana*), but also add a purple-leaved plant, such as *Penstemon* 'Husker Red' or Palace Purple coral bells for contrast.

Cultivars:

'Six Hills Giant' grows to 2 to 3 feet tall with ½-inch, lavender-blue flowers.

Problems: None serious.

Propagation: Division and cuttings. (Seed for *N. mussinii*)

Zones: 3-8

Related Species:

N. mussinii 'Blue Wonder' is a 10- to 12-inch, gray-leaved, blue, seed-propagated catmint. 'White Wonder' is white and 16 inches tall.

N. sibirica is a delightful 2 to 3 foot long flowering blue-violet upright plant that spreads nicely to fill in a garden spot.

N. sibirica 'Blue Beauty' ('Souvenior D'Andre Chaldron') is 18 inches and long flowering, with larger blue-violet flowers.

N. subsessilis has 2-inch, trumpet-shaped, violet-blue flowers, larger leaves, and grows 2 feet tall.

Rudbeckia fulgida

var. sullivantii 'Goldsturm'

Black-Eyed Susan
Asteraceae (Sunflower Family)
○, ◐ Midborder, Mid to late summer

Flowers: These 2- to 2½-inch daisy-like flower heads are golden yellow with a dark brown to reddish brown center.

Leaves: The leaves of black-eyed Susan are a very uniform 2 to 4 inches long. The toothed, pointed, ovate leaves are alternate, dark green, medium thick, and disease-free. The leaves and stems have light hispid (bristly) pubescence.

Growing Needs: This plant does well in light shade to full sun. The soil should be moist but well drained and moderately fertile. Mulching is beneficial.

Landscape Use:

Height: 18 to 24 inches
Habit: Mounded
Spacing: 18 to 24 inches
Texture: Medium

Black-eyed Susan makes an excellent addition to the sunny midborder. Use in large masses, in small groupings, or as a specimen planting.

Design Ideas: Black-eyed Susan combines beautifully with ornamental grasses of all kinds. It also looks great with *Veronica* 'Sunny Border Blue' and hardy geraniums. Grow them with blue asters and white fall anemones.

Varieties:

var. *deamii* has bolder pubescent toothed leaves, is floriferous, and 2 feet tall.

var. *speciosa* is orange flowering with entire leaves (not toothed) and grows to 3 feet tall. 'Viette's Little Suzy' is 12 inches with large golden flowers.

Problems: None serious.

Propagation: Division and self sowing.

Zones: 3-9

Related Species:

The common black-eyed Susan, *Rudbeckia hirta* var. *pulcherrima*, has a thicker, rougher leaf and is usually covered with rust and mildew by midsummer. It is not reliably perennial.

'Irish Eyes' has golden yellow flowers with prominent green centers, 2 to 3 feet tall.

R. nitida 'Autumn Sun' ('Herbstonne') is 5 to 6 feet tall with yellow, down turned petals with a small green cone and handsome dark green glabrous leaves.

'Goldquelle' is double yellow and 2 feet tall.

Salvia × sylvestris

(and nemorosa Cultivars)
(S. × superba)

Perennial Blue Salvia
Lamiaceae (Mint Family)
○ (◗)   Midborder, Summer

Flowers: The flowers of perennial blue salvia are borne on 6- to 8-inch, thin spike-like racemes. Flower color is a violet blue and will fade to gray green with age. Remove faded flowers for a sparse reflowering.

Leaves: The foliage is gray green and leathery with a noticeable midrib. The 3- to 5-inch leaves are opposite and sessile (or clasping) on the stem and pointed lanceolate with crenate (scalloped) margins. The basal rosette leaves are petioled.

Growing Needs: A full sun location will best suit this plant. Perennial blue salvia needs a well-drained soil with moderate fertilization, low in nitrogen. This plant is drought tolerant.

Landscape Use:

Height: 2 to 2½ feet
Habit: Upright mounded
Spacing: 18 inches
Texture: Medium

This wonderful plant is excellent in the sunny mid-border and provides needed flowering transition from May to June.

Design Ideas: Group perennial blue salvia with moonshine yarrow and red poppies. Or combine it with *Iris* 'Stepping Out' (a bicolor of purple and white). It also looks beautiful when planted with *Coreopsis* 'Moonbeam' and pinks.

Cultivars and Hybrids:

'Blue Hill' ('Blauhugel' or 'Blue Mound') has light to middle blue flowers and is long flowering, 20 inches tall.

'Blue Queen' has blue-violet flowers, 18 to 24 inches tall.

'East Friesland' has purple flowers and is 2½ feet tall.

'Lubeca' is long-flowering, violet-blue, with a compact habit, and 18 to 24 inches tall. ('East Friesland' and 'Lubeca' are *S. nemorosa.)*

'May Night' is earlier flowering (late May) with deep blue to purple spikes and has a 18-inch sturdy habit. 1997 Perennial Plant of the Year

'Plumosa' is 2 to 3 feet tall with fuller, thicker, long-flowering, purple flower spikes.

'Rose Queen' is a 2-foot, rose pink flowering selection.

'Rose Wine' is a rose pink version of 'May Night' and is 18 inches tall.

'Snow Hill' is a white version of 'Blue Hill' and grows 18 inches tall.

Problems: White fly (or scale).

Propagation: Cuttings and division.

Zones: 3-8

Related Species:

S. argentea, Silver Sage, is a showy plant with large oval, woolly, silvery-white leaves in a basal rosette. The showy panicles are white (or blushed with pink), 2 feet, and attractive to hummingbirds. Can be short-lived but worth it! I remove spent flowers to prolong its life.

S. azurea ssp. *pitcherii* var. *grandiflora (S. pitcheri),* Blue Sage, has blue flowers, gray-green foliage, and is 3 to 4 feet tall.

S. pratensis (haematodes) 'Haematodes', Meadow Sage, has lilac blue flowers and may be short-lived but will self sow, 2 feet tall.

S. verticillata 'Purple Rain' has large, wavy leaves and full velvety purple spiky flowers and grows 3 feet tall.

Saponaria ocymoides

Soapwort
Caryophyllaceae (Pink Family)
○, ◑   Edging, Summer

Flowers: These ¼-inch, rosy pink, five-petalled flowers are borne in loose cymes on long trailing stems. The calyx is noticeably reddish and glandular-hairy. Buds appear reddish before opening. This plant is just covered with blossoms in early summer.

Leaves: The leaves of soapwort are 1 inch, lanceolate, and opposite on the stem. The stems are reddish in color.

Growing Needs: Plant soapwort in a full sun location. It does best in a very well-drained soil, even sandy and poor soils. Shear back after flowering for more compact vegetative growth and to discourage self sowing.

Landscape Use:

Height:	6 to 10 inches (2 feet spread)
Habit:	Low mounded
Spacing:	2 feet
Texture:	Medium

The low-spreading habit of soapwort makes this an excellent edging or rock garden plant. It also drapes well over walls.

Design Ideas: Soapwort makes a beautiful combination with purple-leaved coral bells. It follows the flowering of creeping phlox and gives a mounded edging effect when planted with columbines, perennial blue salvias, and perennial flax.

Cultivars:

'Alba' has white flowers.
'Carnea' has light pink flowers.
'Rosea' has rose pink flowers.
'Rubra' has rich red flowers.

'Rubra Compacta' has deep red flowers with a more compact habit.
'Splendens' has large deep pink flowers.
'Versicolor' has rose pink buds that open white.

Problems: None serious.

Propagation: Division and seed. Will self seed vigorously.

Zones: 3-7

Related Species:

S. officinalis, Bouncing Bet, is a tall 2 to 3 foot, pink soapwort with a loose open habit. 'Dazzler' has variegated leaves.

Double flowered cultivars — 'Alba Plena', 'Rosea Plena', 'Rubra Plena'

S. × lambergii 'Max Frei' is 1-inch wide, pink, summer-flowering, and grows 12 to 15 inches high.

Scabiosa caucasica

Pincushion Flower
Dipsacaceae (Teasel Family)
○, ◑   Foreground, Midborder, Summer

Flowers: Pincushion flower has a tufted central cushion of disk flowers; hence, the name pincushion. Around this cushion of disk flowers are layers of ragged-edged ray flowers. Each of the 3-inch flowers arise on long stems and come in blue, lavender, white, or pink.

Leaves: The stem leaves are opposite, pinnately divided, and glaucous. The lower basal leaves are linear to lanceolate, and up to 6 inches.

Growing Needs: This plant enjoys full sun to filtered shade but must have well-drained soil. Mulch pincushion flower to keep it moist in summer and to protect it in winter.

Landscape Use:

Height:	1½ to 2 feet
Habit:	Softly mounded to two tiered
Spacing:	12 inches
Texture:	Medium

Use pincushion flower in the sunny midborder of the old-fashioned garden. It also makes an excellent addition to the cut flower garden. Use it massed in one color.

Design Ideas: Plant pincushion flower with evergreen candytuft and penstemon or with soapwort as an edging. Use it as a foreground plant to shrub roses.

Cultivars:

White —

'Alba', 18 inches; 'Bressingham White', 3 feet; 'Miss Wilmott', 3 feet; 'Perfecta' (fringed), 2 to 3 feet; 'Perfecta Alba', 15 inches.

Blue with large flowers—

'Compliment', 15 inches; 'Fama', 18 inches; 'Moerheim Blue' (darker), 2 to 3 feet; 'Penhill Blue' (tinged with mauve), 2 feet.

Lavender Blue or Mixtures—

'Clive Greaves', 'Blue Perfection', 'Isaac House's Hybrids' — mixture (lavender, blue, and white)

Related Species:

S. columbaria 'Butterfly Blue' and 'Pink Mist' (orchid pink) are both 15 to 20 inches tall and very long flowering. Excellent foreground plants!

Problems: None serious.

Propagation: Seed and division.

Zones: 3-7

Related Genus:

Knautia macedonica, Knautia, has burgundy pincushion flowers on 2- to 3-foot stems above tufts of pubescent green entire or pinnatifid foliage. It has a two-tiered habit and thrives in sunny, well-drained areas.

Stokesia laevis

Stokes' Aster
Asteraceae (Sunflower Family)
○, ◑   Foreground, Summer

Flowers: The ray flowers of Stokes' aster are feathery and deeply notched. The solitary 2- to 4-inch flower heads come in lavender-blue or white.

Leaves: The waxy, 4- to 6-inch, pointed, lanceolate leaves of Stokes' aster have a shiny surface and obvious midrib. The upper leaves clasp a tomentose stem.

Growing Needs: This plant is best suited to a sunny, moist, well-drained location but will also withstand a partially shaded location and is somewhat drought resistant.

Landscape Use:

Height:	12 to 18 inches
Habit:	Mounded
Spacing:	12 inches
Texture:	Medium

Stokes' aster makes an excellent specimen planting when used in small groups. It fits well in the sunny foreground of the perennial border.

Design Ideas: Plant Stokes' aster with coral bells, pinks, and *Erigeron*. Another combination with Stokes' aster is Siberian iris, perennial flax, and tunic flower (*Petrorhagia*). A trio of blue *Stokesia*, silver sage (*Salvia argentea*), and pearl yarrow have been at home for years in my garden.

Cultivars:

White—'Alba' and 'Silver Moon'

Lavender Blue—'Blue Danube' is very popular with large flowers (4 to 5 inches wide). 'Blue Moon'

Light Blue— 'Blue Star' and 'Klaus Jelitto'

Darkest Blue—'Wyoming'

Problems: None serious.

Propagation: Division in spring, root cuttings, and seed.

Zones: 5-9

Tanacetum coccineum

(Chrysanthemum coccineum, Pyrethrum roseum)

Painted Daisy, Pyrethrum Daisy
Asteraceae (Sunflower Family)
○, ◑ Midborder, Early summer

Flowers: This daisy-like flower head has a single to double row of ray flowers, which can be rose, red, pink, and white. Each solitary 2- to 3-inch flower head has a yellow center of disk flowers (or may be anemone-like).

Leaves: The 3- to 5-inch foliage of painted daisy is finely divided and fern-like. Each leaf is glabrous, dark green, and alternate on the stem. Painted daisy foliage lacks a pungent odor (as compared to the fern-like and pubescent foliage of the yarrows).

Growing Needs: Painted daisy does well in the full sun or partially shaded (in the afternoon) border. It prefers a well-drained, fertile soil, high in organic matter. Taller varieties may need staking. Mulching is recommended.

Landscape Use:

Height:	24 to 36 inches
Habit:	Upright mounded
Spacing:	18 to 24 inches
Texture:	Medium
	medium fine (in foliage)

Plant this midborder plant in a mass of one color. Or, use it in the cut flower garden. It adds a nice accent to the herb garden as well.

Design Ideas: Combine painted daisy with late spring bulbs. Try grouping it with sea pink as edging, blue camass, and columbines. Position it with Siberian catmint and lamb's ear.

Cultivars:

'Crimson Giant' has large single red flowers, 3 feet tall.

'Eileen M. Robinson' has single salmon pink flowers, 2½ feet.

'James Kelway' has vibrant single scarlet flowers, 2 feet.

'Robinson's Crimson' has single crimson red flowers, 2 feet.

'Robinson's Pink' has single pink flowers, 2 feet.

Mixtures—

Double Mixed Improved

Thompson & Morgan Superb Mixture (semi-double and single in white, pink, and red)

Problems: None serious.

Propagation: Seed and division.

Zones: 3-7

Yucca filamentosa

Spanish Bayonet, Adam's Needle, Yucca
Agavaceae (Agave Family)
○ Background, Summer

Flowers: The 2-inch, ivory flowers of Spanish bayonet are bell-shaped, pendulous, and borne in a large panicle. The flowers of this plant are very fragrant.

Leaves: The sword-like leaves radiate around the plant. Each leaf is rigid, 1½ to 2½ feet long, and has long, curly threads hanging from the tip and margins.

Growing Needs: This easy-to-grow plant enjoys a full sun site in any well-drained soil. Spanish bayonet is difficult to transplant once it is established due to its deep root system. Its outstanding feature is its drought tolerance.

Landscape Use:

Height:	4 to 6 feet in flower
	2 to 3 feet in foliage
Habit:	Spiky mound
Spacing:	3 feet
Texture:	Bold

For a large site, Spanish bayonet makes an excellent specimen planting in the sunny background or center of an island bed. Use it massed (three to five) in a large border.

Design Ideas: Combine yucca with red hot poker. Plant it with plants that have softer texture, such as yarrow, painted daisy, or Russian sage to help soften its boldness.

Cultivars:

Variegated ones (softens texture):

'Bright Edge' — golden yellow leaf edge.

'Bright Eye' — bright yellow margins.

'Golden Sword' — margin is green and center is
    yellow.

'Variegata' — white margins (tinged pink).

Problems: None serious.

Propagation: Division of off shoots.

Zones: 5-10

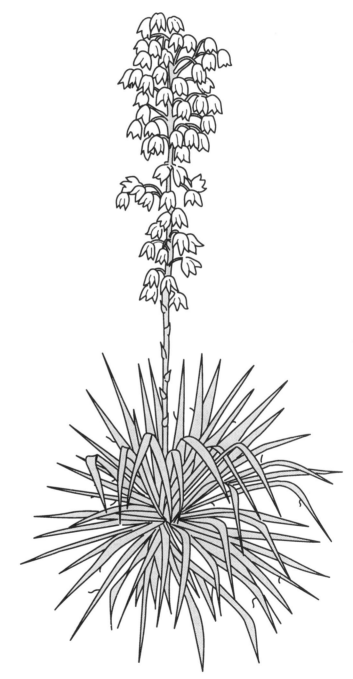

Other Summer Perennials

Achillea tomentosa — Woolly Yarrow

 Sunny foreground to midborder.

 Bright deep yellow flowers.

 Woolly, gray-green leaves.

Aegopodium podograria 'Variegata' — Silveredge Bishop's Weed (Goutweed)

 Very aggressive shady ground cover.

 White umbels, variegated compound leaf, 1 foot, mounded.

 Plant in hard-to-grow areas only.

Anthemis tinctoria — Golden Chamomile or Marguerite

 Sunny, midborder.

 Yellow daisies, aromatic ferny foliage, 2 to 3 feet, mounded, drought resistant.

Aruncus dioicus — Goat's Beard

 Partial shade, moist, background/specimen.

 White fluffy spikes, large astilbe-like foliage, 4 to 5 feet, upright mounded.

Asteromoea mongolica (Kalmia pinnatifida) — Double Japanese Aster

 Partially shaded to nearly full sun, foreground to midborder.

 Small white semi-double flowers, 2 feet, upright airy to mounded.

Bupthalmum salicifolium — Willowleaf Oxeye

 Full sun to partial shade, foreground to midborder.

 Yellow daisy, attractive linear foliage, 2 feet, mounded.

Calamintha grandiflora (Large-flowered Calamint) ('Variegata')
C. nepeta (Calamint)

 C. grandiflora — 12 to 15 inches.

 C. nepeta — 18 to 24 inches.

 Full sun, well-drained, foreground.

 Lavender two-lipped flowers (*nepeta* in whorls, *grandiflora* larger, cymose), aromatic, ovate, scalloped leaves, mounded.

Clematis hybrids — Clematis

 Sunny vine (shade the roots).

 Large, showy flowers in a range of colors, leaflets usually in threes, woody or semi-woody, 6 to 12 feet vining.

Coreopsis auriculata — Mouse Ear Coreopsis

 Full sun, well drained (native).

 Dark yellow, single flower, toothed edges.

 Low growing, 6 to 8 inches and long flowering, two-tiered, foreground.

Digitalis grandiflora — Large Yellow Foxglove

 Long yellow racemes, partial shade, midborder, 2 to 3 feet.

D. × mertonensis — Strawberry Foxglove

 Strawberry pink racemes, partial shade, background, 3 to 4 feet.

D. purpurea — Foxglove (Biennial)

 Moist, partial shade, midborder to background, 2 to 4 feet.

 Showy one-sided, spotted racemes in purple, pink, white, or yellow.

 Allow seeds to self sow.

 'Excelsior Hybrids', 'Foxy' — flowers are borne all around the raceme.

Euphorbia griffithii 'Fireglow' — Griffith's Spurge

Orange-red bracts, full sun to partial shade, attractive foliage, fall color, mounded, midborder, 2 to 3 feet.

Filipendula vulgaris — Dropwort Meadowsweet

Full sun to partial shade.

White fluffy terminal panicles, upright mounded, midborder, 2 feet.

F. rubra — Queen-of-the-Prairie

Full sun, moist.

Native plant, pink fluffy panicles, upright, two-tiered, 6 to 8 feet.

Gaura lindheimeri — White Gaura

Full sun, well drained.

White airy spikes (pink cultivars), small foliage, upright airy, midborder, 3 to 5 feet.

Helianthemum nummularium — Sunrose

Sunny, well-drained, alkaline, sandy loam soils. Mounded, foreground, 1 foot.

Small 1-inch flowers in yellow, orange, red, pink, white, bicolors.
Small gray-green foliage.

Houttuynia cordata 'Chameleon' — Chameleon Houttuynia

Full sun to partial shade.

Heart-shaped leaves in red, ivory, and green. Small flower spike.

Confined edging/ ground cover/containers, 8 to 10 inches.

Beware of rapid growth in moist sites.

Inula ensifolia —Inula

Full sun, light shade.

Yellow daisies, neat lanceolate foliage, mounded, 1 foot, edging, foreground.

Lathyrus latifolius — Perennial Sweet Pea

Sunny 6 to 8 foot vine.
Pink, purple, white.

Lobelia cardinalis — Cardinal Flower

Partial shade, moist.

Showy scarlet red racemes and burgundy foliage, upright, 3 to 4 feet, background.

Liriope spicata — Creeping Lilyturf

Partial shade (sun).

Ground cover, 1 foot.

Linear grass-like foliage, thin violet or white racemes, blue-black berries.

Liriope muscari — showier, but not hardy.

Lunaria annua — Money Plant, Silver Dollar Plant (Biennial)

Great seed pods.

Full sun, partial shade, well drained.

Pink to purple or white racemes, upright mounded, 2 to 3 feet, midborder.

Malva alcea 'Fastigiata'— Hollyhock Mallow

Full sun, partial shade, well drained.

Lobed foliage, 3 to 4 feet, pink notched hollyhock-like flowers.

Upright, midborder, background.

Oenothera missouriensis — Missouri Primrose, Ozark Sundrops

Yellow, full sun, well-drained edging to foreground.

O. speciosa — Showy Primrose

Pink, pinkish-white flowers.

Full sun, well drained, infertile, foreground, 1 to 2 feet.

Can be quite aggressive

Penstemon barbatus —
Penstemon or Beardtongue

1½ to 3 feet, pink/blue/white.

Green foliage.

Partial shade to full sun, foreground to midborder.

P. digitalis 'Husker Red'

White flowers, burgundy red foliage, 2½ to 3 feet.

Partial shade to full sun.

Moist, well drained, upright mounded, midborder.

Flower shape is a combination of coral bells and foxglove.

1996 Perennial Plant of the Year.

Petrorhagia saxifraga (Tunica saxifraga) —
Tunic flower

Sunny, moist, well drained, edging.

Long-flowering baby's breath-like flowers and wispy foliage, 6 to 10 inches.

Sagina subulata — Pearlwort, Irish Moss

Full sun, partial shade, well drained.

White flower, prostrate, 2 to 4 inches, edging.

Sanguisorba obtusa — Japanese Burnet

Full sun, moist, well drained, mulch.

Fluffy pink terminal spikes.

Nice, medium-green, compound foliage.

Midborder to background, upright mounded, 3 to 4 feet.

Sedum kamtschaticum — Kamtschat,
Sedum or Stonecrop

Full sun, well drained.

Drought resistant, low mounded, 4 to 8 inches.

Yellow starry flowers, turning orange, then brown.

Whorled-looking (alternate) foliage, edging or ground cover.

'Variegatum' — green, white, and pink.

Sempervivum tectorum —
Hen and Chickens

Full sun, well drained.

Attractive whorled rosettes of pointed, green and reddish, or purplish leaves.

Unique tall, pubescent flowers (which I usually remove).

Edging, 2 to 4 inches (12 to 15 inches in flower).

Sidalcea malviflora — Miniature Hollyhock

Full sun or partial shade.

Moist, light to dark pink, purple, red, white.

Rounded foliage.

Upright, midborder, 2 to 4 feet.

Silene dioica — Red Campion

Full sun, well drained.

Reddish purple cymes, single flowers with unique inflated calyx.

Soft pubescent gray-green foliage, upright to mounded, midborder, 2 to 3 feet.

'Robin's White Breast' — white, 6 to 8 inches.

'Swan Lake' — double white, 6 to 8 inches.

Tanacetum parthenium — Feverfew

(Chrysanthemum or Matricaria)

Full sun, well drained.

Small white, single to double flowers.

Aromatic, fern-like foliage, 1 to 3 feet.

Mounded, foreground to midborder.

'Aureum' — golden foliage and single flowers, 1 foot.

'Snowball' — double white flowers for fresh or drying, 2 to 3 feet.

Self-sowing biennial.

BULBS

Allium giganteum

Giant Allium
Liliaceae (Lily Family)
○ Midborder, Background, Early summer

Flowers: These small, bright lavender flowers are borne in large globose umbels, 4 to 5 inches across, on a 4 to 5 foot leafless scape.

Leaves: The leaves are 12 to 18 inches long and 2 to 3 inches wide. The leaves are linear and form a basal rosette. The leaves emerge early and are often damaged by frost causing browning along the edges.

Growing Needs: Plant this bulb in fall in a sunny, well-drained location. Site it in a wind-sheltered site. Mulch is recommended. Allow the foliage to yellow before removing.

Landscape Use:

Height:	4 to 5 feet
Habit:	Upright lollipop
Planting depth:	6 inches
Spacing:	12 inches
Planting time:	Fall
Texture:	Medium bold to bold

Plant giant allium in the sunny background of the border. Placement in front of shrubs helps give it a good background and some wind protection. It makes a fun addition to a children's garden or the cut flower garden.

Design Ideas: Giant allium looks wonderful when interplanted with tickseed and silver king artemisia. An evergreen shrub border really sets off the flowers of giant allium. Use this plant in groups of 8 to 10 massed as an accent near Alaska Shasta daisies and delphiniums.

Problems: None serious. (Bulb rot—select healthy bulbs)

Propagation: Division and seed (slow).

Zones: 5-8

Related Species:

Allium christophii, Star-of-Persia, is 18 inches to 2 feet high and produces metallic lavender umbels that are 8 to 12 inches across. This plant makes an excellent specimen planting of three to five in the foreground next to dill 'Bouquet' with snow in summer as edging. It is beautiful arising out of *Geranium striatum* or other hardy geraniums.

Allium sphaerocephalum, Drumstick allium, is 18 inches high and produces an oval-shaped umbel that is 1 to 2 inches across and 2 to 3 inches long. The petals are burgundy to dark lavender in color. The linear foliage emerges in fall and overwinters.

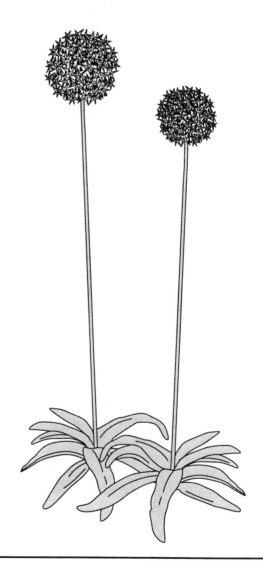

Lilium

Lily
Liliaceae (Lily Family)
○ Midborder, Background, Summer
(early, mid, late season)

Flowers: Each trumpet- to star-shaped flower is composed of three petals and three petal-like sepals. The 3- to 6-inch flowers come in many colors, such as cream, white, yellow, orange, red, pink, lavender, and bicolors. The blooms will face either outward, upward, or downward.

Leaves: Each narrow lanceolate leaf is 3 to 5 inches long. The glossy leaves are whorled around the stem, often appearing crowded.

Growing Needs: Lilies should be grown in the sunny midborder to background. Midday shade will extend the flowering period. Plant the bulb in fall in a well-drained site. Raised beds with added organic matter will reward you with high-quality lily flowers and plants. Taller varieties should be staked.

Landscape Use:

Height:	2 to 6 feet
Habit:	Upright, Vase shaped
Planting depth:	6 inches
Spacing:	6 inches
Planting time:	Fall
Texture:	Medium

Use lilies in large groupings or as a specimen planting of just three to five in the midborder or background of the garden. No garden should be without the regal, royal lily.

Design Ideas: Plant lilies with clustered bell-flowers, balloon flowers, and yarrows. Interplant taller pink varieties in silver king artemisia.

Types:

Asiatic — early summer, 1 to 3 feet, upward facing flowers. 'Enchantment', 'Connecticut King'

American — mid summer, 2 to 3 feet, outward facing flowers, some staking.

Aurelian — mid summer, 4 to 5 feet, downward or outward trumpets, 'Regale', 'Golden Splendor'.

Oriental — late summer, 4 to 5 feet, outward or downward facing, 'Casa Blanca', 'Stargazer' (faces upward).

Problems: Lily mosaic (virus transmitted by aphids), basal bulb rot (destroys or damages bulbs in poorly drained soils).

Propagation: Division, scales (remove outer ones from the mother bulb), seed, bulbils.

Zones: 4-8

Other Summer Bulbs

Allium karataviense — **Turkestan Onion**

Full sun, well drained.

Lavender globose umbel, 4 to 6 inches wide.

Two wide (4 inches across) leaves, 8 to 10 inches long.

Mounded, specimen, foreground, 10 inches.

Allium senescens — **Flowering Onion**

Full sun, well drained.

Rose-lavender, 3-inch, rounded umbels.

Flat basal foliage.

Upright mounded, 2 feet.

Foreground, midborder.

A. s. var. glaucum

Full sun, well drained, curly, swirling blue-green foliage.

Lilac flowers.

Mounded, foreground to edging, 1 foot.

Eremurus × *isabellinus* — **Foxtail Lily**

Full sun, well drained.

Showy erect dense racemes in white, pinks, yellows, orange to red.

Upright, background, 4 to 6 feet. Stake.

MID TO LATE SUMMER and FALL PERENNIALS

Aconitum napellus

Monkshood, Helmet Flower
Ranunculaceae (Buttercup Family)
◑ Midborder, Background, Mid to late summer

Flowers: Monkshood produces flowers that are helmet- or hood-like in shape and borne on 4- to 6-inch racemes. Each 1- to 1½-inch flower is blue, purple, white, or bicolored.

Leaves: The attractive foliage is alternate, palmately divided, and deeply dissected. Each 2- to 4-inch leaf is dark green and glossy with no pubescence. They emerge early in spring and remain attractive all through the growing season.

Growing Needs: Monkshood does well in a partially shaded location. A moist and moderately fertile, but well-drained soil is preferred. This plant would benefit from mulching. The plants can remain undisturbed for years. The leaves and roots of monkshood contain aconite, which is poisonous when swallowed. Therefore, do not plant it near a child's play area or a vegetable garden.

Landscape Use:

Height:	3 to 4 feet
Habit:	Upright to upright mounded
Spacing:	18 inches
Texture:	Medium

Plant monkshood in the partially shaded midborder or background of the perennial border for its needed blue color. It also makes an excellent specimen planting and cut flower.

Design Ideas: This plant looks great with Hosta and black snakeroot (*Cimicifuga*) with perennial forget-me-not (*Brunnera*) for a foliage edging. Monkshood combines well with fall anemone and Japanese painted fern.

Cultivars:

'Album' has a white flower.

'Carneum' has a pink flower and fares better in the cooler climates.

Hybrids:

A. x cammarum (bicolor) 'Bicolor' has showy blue and white branching flowers, 3 to 4 feet.

'Blue Sceptre' is 2 feet tall with blue flowers.

'Bressingham Spire' is a sturdy 3 feet tall, violet-blue.

MID TO LATE SUMMER and FALL PERENNIALS

'Eleanor' has large near-white flowers with a blue edge and is 3 feet tall.

'Newry's Blue' has dark blue flowers and is 4 to 5 feet tall.

'Stainless Steel' is a metallic blue that grows 3 feet high.

A. carmichaelii, Azure Monkshood, is later flowering (August to September) and grows 2 to 3 feet tall.

'Arendsii' is azure or deep blue or blue-violet, 3 to 4 feet tall, and sturdy—does not need staking.

'Barker's Variety' is very tall (6 feet), violet (perfect for the autumn garden), with leathery dark green foliage.

'Latecrop' has blue flowers and is later flowering (September).

A. henryi 'Spark' ('Spark's Variety') is a tall cultivar, 4 to 5 feet, with violet-blue, well-branched flowers.

A. lycoctonum ssp. *neapolitanum* (*A. lamarckii*) is yellow and relatively sturdy, 3 feet.

A. orientale is tall (4 to 5 feet), summer flowering, in dark blue or white, flushed yellow.

Problems: None serious.

Propagation: Division (slow to establish) and seed (difficult).

Zones: 3-8

Anemone × hybrida

Fall Anemone, Japanese Anemone
Ranunculaceae (Buttercup Family)
○, ◑ Midborder to background, Late summer
to early fall

Flowers: The 2- to 3-inch, semi-double, white,
pink, or mauve flowers are either single or double
with a showy yellow center. The buds (sepals) are a
darker color, opening lighter. This is one of the
best late season flowers.

Leaves: The attractive foliage of fall anemone is
palmately compound with three ovate, toothed,
and lobed leaflets. Each dark green leaf is 5 to 7
inches across with a very long petiole. The foliage is
also arranged in a whorl beneath each flower inflorescence. Fall anemone foliage emerges early in
spring and is a nice textural presence all summer.

Growing Needs: Japanese anemone prefers a
partial shade location, particularly on the east side
of a home or fence for a morning sun exposure. The
soil should be fertile, moist, well drained, and high
in organic matter. Mulching is beneficial. This
anemone can be quite prolific when grown in a preferred site; it forms a nice colony so give it room.

Landscape Use:

    Height: 2 to 4 feet
    Habit: Mounded or two tiered
    Spacing: 18 to 24 inches
    Texture: Medium

Plant Japanese anemone in mass in the sunny midborder or background of the garden. It is a welcome addition because of its late flowering. It is
an attractive short-lived fresh cut flower.

Design Ideas: Combine Japanese anemone
with 'Purple Dome' asters as a foreground plant,
with white and blue balloon flowers next to it, and
with monkshood and ornamental grasses behind it.
The leaf texture combines beautifully with hardy
geranium (any and all) leaves.

Cultivars:

There is a lot of confusion about correct scientific
naming of many anemone cultivars and hybrids.
The 1992 Royal Horticultural Society *Dictionary of
Gardening* was helpful in sorting these into proper
categories.

'Alba' has white flowers, 2 to 3 feet.

'Coupe d'Argent' has wavy cream flowers, fading
    white, 3 feet.

'Elegantissima' is a double pink, 4 feet.

'Honorine Jobert' is a lovely single white with dark green leaves, 3 to 4 feet.

'Konigin Charlotte' ('Queen Charlotte') is a large (3 inches) semi-double, light pink that grows 3 feet high.

'Kriemhilde' has pale pinkish-purple, semi-double flowers and is 2 to 3 feet tall.

'Lady Gilmore' has large, semi-double, pink flowers and is 3 feet tall.

'Margarete' has a deep pink, semi-double flower, 3 feet.

'Mont Rose' has a rose colored double flower on 2½-feet stems.

'Max Vogel' has large (3 to 4 inches across) single pink flowers and is 3 to 4 feet tall.

'Pamina' is a dark rose pink double anemone that grows nearly 3 feet tall.

'Prince Heinrich' has semi-double purplish-red flowers, 3 feet tall.

'Whirlwind' is a lovely semi-double white that grows 3 to 4 feet tall.

A. hupehensis, Fall Anemone, is usually shorter, has smaller flowers, and tolerates drier, sunnier sites.

'Bowles Pink' has deep pink flowers and is 3 feet tall.

'Hadspen Abundance' is two-toned, purple-pink, single flowering, and 3 feet tall.

'September Charm' has single silvery-pink flowers, opening from dark pink buds. It is a prolific 3-foot high plant.

A. hupehensis var. *japonica*

'Bressingham Glow' has deep rosy pink (almost red) semi-double flowers on 2 to 3 feet compact stems.

'Praecox' is an early, dark pink, single anemone, 2 to 3 feet tall.

'Prinz Heinrich' ('Prince Henry') is a early flowering, rose pink, semi-double, compact 2- to 2½-foot plant.

Anemone tomentosa (vitifolia) is an earlier flowering anemone with white tomentose pubescence on the underside of the leaves.

'Alba' has white flowers and is 3 feet tall.

'Robustissima' is a hardy (Zone 3), early flowering plant with soft pink single flowers, 3 feet.

Problems: None serious.

Propagation: Spring division (easy), fresh ripe seed, root cuttings in winter.

Zones: 4-8

Artemisia schmidtiana

'Silver Mound' or 'Nana'

Silver Mound Artemisia
Asteraceae (Sunflower Family)
○ Edging, Foreground, Mid to late summer

Flowers: The flowers are very small, white, and nearly inconspicuous. This plant is grown for its foliage.

Leaves: This excellent foliage plant has alternate leaves which are crowded along the stem. The 1- to 1½-inch silver leaves are finely divided, appear feathery, and are soft to the touch.

Growing Needs: Silver Mound artemisia thrives in the full sun with well-drained soil. It grows in infertile soil conditions and tolerates drought. If given too much moisture or fertilizer, it will become quite lush and flop open in the center. (Ignore it for best results.)

Landscape Use:

Height:	8 to 12 inches
	('Nana' 3 to 4 inches)
Habit:	Low mounded
Spacing:	12 to 18 inches
Texture:	Fine

Plant Silver Mound artemisia as a sunny edging or foreground plant as an accent or in small groups (two or three). Try repeating a single specimen throughout the border. Silver Mound is quite at home in the rock garden.

Design Ideas: Combine Silver Mound artemisia with low-growing sedums. Try it with gray-green basket of gold leaves, clustered bellflowers, and Russian sage.

Problems: High humidity and heat will cause the foliage to rot and turn black.

Propagation: Stem cuttings.

Zones: 3-8

Related Species:

Artemisia ludoviciana cultivars, white sage, make an excellent accent in the midborder or background of the garden.

'Silver King' is a 3-foot tall plant with 2- to 3-inch, fine textured, lanceolate, silver leaves and a lovely fine texture. I have planted two types called Silver

King. The one with the more fine textured foliage was not invasive, the other one was (the wider the foliage — the more spreading). Place it in a confined or poor, infertile site to watch its spread for a season. It is easy to divide (or spade or pull out). Excellent for fresh or dried arrangements.

'Silver Queen' has jagged, toothed leaves that are ovate to rounded lanceolate. Its texture appears less fine compared to Silver King. It spreads quickly in cultivated garden soil. Place it next to plants that will hold their own, such as Russian sage, Autumn Joy sedum, and yarrows. Divide and remove as necessary. It is 2 to 2½ feet tall.

'Valerie Finnis' has wide (½ inch) silver-gray, jagged, or toothed, leaves on a 2-feet, compact plant. It is a showy blending plant for the perennial border.

Artemisia x 'Powis Castle' is a lovely fine-textured, 2- to 3-feet tall plant with silvery filigreed leaves. Although it rarely flowers, the leaves are a showy attraction in the garden. It is marginally hardy but well worth planting for its rounded habit and ability to withstand heat and humidity.

Artemisia lactiflora, White Mugwort, is a unique, tall (4 to 6 feet), background plant with showy, fluffy ivory panicles (1 to 1½ feet long) and bold green palmately divided foliage, flowering in late summer to early fall. This artemisia has the coarsest leaves of the landscape artemisias. Quite eye-catching!

A. lactiflora 'Guizho' has dark green leaves, purplish-green stems, and cream-colored, astilbe-like flowers in late summer. Versatile for sun or shade, 5 feet tall.

Artemisia stellerana, Beach Wormwood, Perennial Dusty Miller, has nice silky white foliage resembling annual dusty miller. Yellow panicles appear in early summer. Use in the sunny foreground (1 to 2 feet). Somewhat aggressive.

'Silver Brocade' is a very low-growing ground cover with fuller leaves and 1-foot yellow flowers in early summer.

Artemisia absinthium 'Lambrook Silver' is a 2½-foot tall midborder plant with silvery dissected leaves.

'Lambrook Giant' is taller at 3 to 4 feet.

Asclepias tuberosa

Butterfly Flower, Butterfly Weed
Asclepiadaceae (Milkweed Family)
○ Midborder, Mid to late summer

Flowers: This wonderful orange (yellow or red) flower is produced in terminal 2- to 5-inch, flat-topped, umbel-like cymes (resembling a butterfly wing). Each individual flower is ½ inch and five-petalled. Ornamental, linear, pointed seed pods appear in early fall.

Leaves: Each 2- to 4-inch leaf is narrowly lanceolate with hispid pubescence. The leaves are medium green in color and opposite and dussuate (both opposite and alternate, often appearing whorled) on a stem. The milky sap produced when a leaf is removed is an easy identification feature. Another ID feature is the presence and nibbling of the Monarch butterfly caterpillar.

Growing Needs: Butterfly flower does well in a sunny, well-drained site. This plant is drought tolerant and dislikes transplanting. It is very low maintenance. Butterfly flower is excellent for fresh flower designs or for dry flower arrangements once the seed pods have formed.

Landscape Use:

Height:	1½ to 3 feet
Habit:	Vase shaped
Spacing:	18 to 24 inches
Texture:	Medium

Butterfly weed looks best if massed or naturalized in a sunny spot. Its striking orange color also makes it a good specimen plant.

Design Ideas: For an excellent meadow garden, combine it with blazing star, purple coneflower, blackberry lily, and with white Jupiter's beard. Combine it with *Veronica* 'Sunny Border Blue' for contrast.

Cultivars:

'Gay Butterflies' has red, orange, and yellow flowers in many shades, 2 to 2½ feet.

Problems: None serious.

Propagation: Fresh seed, root cuttings, stem cuttings, careful division because of long taproot. Buy small containerized plants because of its taproot.

Zones: 4-9

Related Species:

A. incarnata, Swamp Milkweed, is tall (3 to 4 feet), a lovely deep pink color, and ideal for moist sun or shade.

'Ice Ballet' is white, 3 to 4 feet tall.

Aster × frikartii

Frikart's Aster
Asteraceae (Sunflower Family)
○, ◑   Foreground, Midborder, Mid to late
summer to fall

Flowers: Frikart's aster produces a multitude of daisy-like, violet-blue flower heads with yellow disk flowers in the center. These fragrant, 2½-inch flowers are great for the cut flower garden.

Leaves: Each 1- to 3-inch leaf is lanceolate to oblong in shape, sessile, and alternate on the stem. These gray green leaves and stems are pubescent. The leaf margin is entire with some serrations at the tip. Generally, Frikart's aster is mildew resistant.

Growing Needs: Site Frikart's aster in a sunny location with well-drained soil. Mulch for winter survival where there is no snow cover. Likewise, do not trim back the over-wintering foliage until spring. Taller plants may require peastaking.

Landscape Use:

Height:	2 to 3 feet
Habit:	Mounded
Spacing:	18 inches
Texture:	Medium

Frikart's aster makes an excellent addition to the sunny foreground as a specimen or massed. It is also good for masking foliage of other plants that may die back or become unsightly.

Design Ideas: Plant in front of fading tulip foliage, along with lilies, white obedient plant, and Russian sage. Combine with maiden pinks and coral bells as the edging and foreground plants. Frikart's aster is an excellent blending plant to soften the textures and colors of other plants nearby.

Cultivars:

'Jungfrau' has deep lavender blue flowers and is a
    compact 2 feet tall.

'Monch' is an outstanding 2 to 3 feet tall plant with long-flowering, large, lavender blue flowers. It has a sturdy mounded habit and is well branched.

'Wonder of Stafa' has lavender blue flowers on 2½- to 3-feet stems.

Problems: None serious.

Propagation: Division.

Zones: 5-9 (Site well in Zone 5 — on the east side of a house out of the wind and provide winter mulch.)

Related Species:

For later flowering asters, try *Aster novae-angliae*, New England Aster or *Aster novae-belgii*, New York Aster. Numerous cultivars!

Aster novae-angliae, New England Aster or Michaelmas Daisy — This native species thrives in full sun and well-drained soils. These asters flower late, usually throughout the fall, and can be pinched back in June to encourage more compact growth. The color range is white, pink, nearly red, blue, and purple. The flowers are 1 to 2 inches wide borne in corymbose panicles. The gray-green, medium-textured leaves are alternate, sessile, lanceolate, entire, pubescent, and 3 to 5 inches long. New England asters range from 3 to 6 feet with upright mounded to mounded habits for the background. Cut back faded flowers to prevent self sowing.

'Alma Potschke' has vibrant hot pink flowers and grows 3 feet tall.

'Purple Dome' is completely covered with dark purple flowers on a mounded, 18-inch plant. Outstanding!

'Harrington's Pink' is an older cultivar that has pink flowers, 4 feet.

'Hella Lacey' is a tall (4 to 5 feet) plant with large, 2+ inch flowers. It needs support.

'Honey Song Pink' has bright pink flowers and grows 3 to 4 feet tall.

'September Ruby' has rosy red flowers and are 4 feet tall.

'Treasure' has very large bright purple flowers and is 3½ to 4 feet high.

'Wedding Lace' is tall (5 feet) and white.

Aster novi-belgii, New York Aster or Michaelmas Daisy — This native species also needs full sun and well-drained soils. These late flowering asters are shorter (1 to 3 feet) than New England asters and are mounded foreground or midborder plants of medium texture. The 1½- to 2-inch flowers are borne in corymbose panicles. The alternate, lanceolate leaves are medium green, 3 to 5 inches long, and pubescent. Pinch back the plants at 6 to 8 inches high to encourage compact growth.

'Alert' has red flowers, 12 to 16 inches.

'Crimson Brocade' is a semi-double, crimson red, 30 inches.

'Fellowship' has very large double pink flowers and is 3 feet high.

'Jenny' is raspberry red and 2 to 3 feet.

'Marie Ballard' is a double pink, 3 feet tall.

'Patricia Ballard' is 3 feet and rose-pink.

'Professor Kippenberg' is 15 to 18 inches, blue, with a yellow center.

'Red Star' is 12 to 15 inches, rose red.

'Snow Flurry' is 18 inches, large, and white.

A. cordifolius 'Little Carlow' is 3 feet, blue, and ideal for September and October.

A. divaricatus is 18 inches, small, white, and suitable for partial shade.

A. ericoides 'Monte Cassino' is fine textured, 18 inches, and white.

A. lateriflorus 'Coombe Fishacre' is rose pink with a 2-foot, upright habit.

A. laevish, Smooth Aster, is lavender or blue, 3 to 4 feet tall, and very dependable and drought resistant.

A. l. var. *horizontalis*, Calico Aster, is 2 feet, lilac-colored, with attractive small foliage (orange fall color).

Belamcanda chinensis

(Pardanthus chinensis)

Blackberry Lily
Iridaceae (Iris Family)
○, ◑ Midborder, Mid to late summer

Flowers: Each 1- to 2-inch flower is 6-petalled, star-shaped, and orange flecked with red. Each flower lasts for one day only, twists in a spiral as it fades, and eventually produces a fruit that resembles a ripe blackberry. Many flowers are borne on a branched stem so flowering extends for quite a while.

Fruits: The capsule splits to reveal glossy, round, black seeds that resemble blackberries. The seeds persist and will self sow. To use for dried arrangements, harvest the stem when the capsule changes from green to tan, but before it splits opens.

Leaves: The leaves are 1 inch wide, sword-shaped, and 10 to 12 inches long. The leaves look like iris foliage but are medium green to yellow-green, not blue-green. Its stair-stepped, fan-shaped arrangement will be taller than iris. (Iris foliage arises from the rhizome.)

Growing Needs: Blackberry lily needs a full-sun site and well-drained soil. More lush growth and height is produced in moist, fertile soils.

Landscape Use:

Height: 2 to 3 feet in flower
Habit: Vertical, vase shaped
Spacing: 12 inches
Texture: Medium

Plant blackberry lily in the sunny midborder, massed or as a specimen planting. Its vase-shaped habit contrasts well with mounded habits of other plants. The seed pods are delightful in dried flower arrangements.

Design Ideas: Combine orange blackberry lily with yellow Kamschatka sedum as an edging and with Goldsturm black-eyed Susan in the background. Hello Yellow also combines well with hardy geraniums, blue or purple asters, and garden mums.

Cultivars:

'Hello Yellow' is a lovely solid yellow.
'Freckle Face' is pale orange, 2 feet.

Problems: Iris borer. Practice prevention by removing and destroying the leaves in winter.

Propagation: Seed (harvest the fresh seed), transplant small self-sown seedlings, and division.

Zones: 5-10

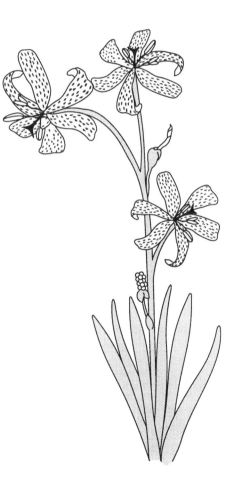

Boltonia asteroides

Boltonia
Asteraceae (Sunflower Family)
○   Background, Late summer to fall

Flowers: These 1-inch, daisy-like flower heads have white (pink or lavender) ray flowers and yellow disk flowers. The flower heads are borne profusely in loose panicles.

Leaves: Each 3- to 5-inch, willow-like leaf is linear to lanceolate, gray green, and glaucous. The leaves are sessile alternate on the stem. Boltonia has very sturdy stems that are well branched.

Growing Needs: Boltonia prefers full sun and a moist well-drained soil. It is a low-maintenance plant; just trim it back in the early spring. Boltonia will tolerate light shade but may need staking because of it. The cultivars are sturdy plants that do not require staking.

Landscape Use:

Height:	4 to 6 feet
Habit:	Upright to upright mounded
Spacing:	18 inches to 2 feet
Texture:	Medium fine

Boltonia cultivars are a must for the sunny background of the fall border.

Design Ideas: Combine Snowbank boltonia with Japanese anemone, Joe-pye weed, and obedient plant. Another great combination for *Boltonia* 'Pink Beauty' is with Alma Potschke aster and Russian sage with Autumn Joy sedum as a foreground plant and blue plumbago (*Ceratostigma*) as edging.

Cultivars:

'Snowbank' has prolific white flowers and is a compact, sturdy 4 feet tall.

'Pink Beauty' is also 4 feet tall and has numerous light pink flowers.

Problems: None serious. Unlike fall asters, it is mildew resistant.

Propagation: Division and cuttings.

Zones: 3-9

Catananche caerulea

Cupid's Dart
Asteraceae (Sunflower Family)
○   Foreground, Mid to late summer

Flowers: The 1- to 2-inch flower head is made up of pinked (toothed) ray flowers, which are blue or white. Each flower head is subtended by papery bracts and borne on long slender peduncles.

Leaves: These long, 7- to 10-inch, narrow leaves are lanceolate and appear gray because of the tomentose hair. The leaves are borne in a basal rosette and sparsely alternate on the stem.

Growing Needs: Cupid's Dart prefers a sunny site with a well-drained soil. Wet winter soils will kill this plant. This short-lived perennial should be replaced every three to five years to achieve maximum flower potential.

Landscape Use:

Height:	15 to 18 inches
Habit:	Upright airy to mounded airy
Spacing:	12 inches
Texture:	Medium fine

Use this plant in the foreground of the sunny perennial border, in rock gardens, or in the cut flower garden. The flowers may be used fresh or dried; the dried papery bract is also very interesting in arrangements.

Design Ideas: Cupid's dart looks great with snow-in-summer as an edging and with hardy geraniums and garden phlox. Use thyme or low-growing sedums as an edging to show off *Catananche* as an accent. Combine blue cupid's darts with pink long-flowering *Centranthus ruber* and silvery Powis Castle artemisia.

Cultivars:

'Alba' has white flowers.

'Bicolor' has white petals with a dark blue center. Showy!

'Blue Giant' has pale blue flowers and silvery leaves.

'Major' has lavender flowers and grows 2½ feet tall.

'Perry's White' has white flowers.

Problems: None serious.

Propagation: Seed and division.

Zones: 4-9

Centaurea macrocephala

Globe Centaurea
Asteraceae (Sunflower Family)
○ Background, Mid to late summer

Propagation: Seed (and self sows), transplant small seedlings, spring or fall division.

Zones: 3-9

Flowers: These bold, 3-inch, yellow flower heads are solitary, globose, feathery, and subtended by brown papery bracts. It looks like a yellow thistle and is quite showy up close and at a distance. Globe centaurea makes an excellent cut flower, fresh or dried.

Leaves: These broad lanceolate leaves can be 3 inches wide and 8 to 10 inches long in the basal rosette, then decreasing in size to the top of the stem in an alternate arrangement. Each leaf is covered with rough (hirsute) hairs and feels like sandpaper. The stems are also pubescent.

Growing Needs: Globe centaurea will thrive in a full sun, well-drained location. This plant can also tolerate drought conditions.

Landscape Use:

Height:	3 to 4 feet
Habit:	Upright mounded
Spacing:	2 feet
Texture:	Bold

The boldness of this plant allows for it to be easily viewed when placed in the background of the perennial, wildflower, or cut flower border. It can placed as a single specimen or a massed planting

Design Ideas: Plant globe centaurea with white feverfew and lady's mantle. Combine it with gooseneck loosestrife, red or orange lilies, yarrow, and yellow corydalis. Six Hills Giant catmint and sea lavender contrast and soften its boldness very effectively.

Problems: None serious.

Ceratostigma plumbaginoides

(Plumbago larpentiae)

Plumbago, Blue Plumbago, Leadwort
Plumbaginaceae (Plumbago Family)
○, ◑   Ground cover, Edging, Late summer to fall

Flowers: The flowers of plumbago are borne in terminal clusters. Each ½-inch, five-parted flower is a rich blue with attractive contrasting red pointed buds. It is valuable for providing blue color late in the growing season.

Leaves: The 1- to 3-inch leaves are alternate, obovate to rhomboid (diamond-shaped), and fringed with hairs along the margin (ciliate). In the fall, the leaves turn a beautiful coppery red.

Growing Needs: The soil must be well drained for plumbago. Place it in full sun or partial shade. Plumbago is late to emerge in spring so do not accidentally disturb it. I leave the over-wintered stems to remember where it is planted and then trim the stems back as the new growth emerges. In colder areas (northern zone 5), plumbago will need a winter mulch or should be sited on the south side of a house in a protected location. After a very cold winter (and no snow cover), the plumbago in my garden (on the west side of my house) may not be as thick, but it has never died out (mid Zone 5). It has lush growth by the time it flowers in August. A great back-to-school plant.

Landscape Use:

Height:	6 to 12 inches
Habit:	Low mounded
Spacing:	12 inches
Texture:	Medium

Plumbago makes an excellent sunny to partially shaded ground cover. Use it as an edging when it has some room to spread or in the rock garden.

Design Ideas: This tough plant looks great planted next to Autumn Joy or Brilliant sedum, white obedient plant, and Sunny Border Blue veronica. It fills in nicely with soapwort along a stepping stone path. It also makes a nice edging near variegated iris, catmint, and purple coneflower.

Problems: None serious.

Propagation: Spring division and cuttings.

Zones: 5-9

Chryanthemum × morifolium

(Dendranthema × grandiflorum)

Garden Mum, Hardy Chrysanthemum
Asteraceae (Sunflower Family)
○ Foreground, Midborder, Late summer until frost

Flowers: The variety of types, sizes, and colors of garden chrysanthemums is tremendous. The 1- to 6-inch flower heads are available in a range of colors and shades, such as white, pink, lavender, yellow, orange, bronze, and red. The flower types include daisy or single, anemone, semi-double, buttons, pompons (ball-shaped), or double. The petals (ray florets) can be strappy or shaped like spoons, quills, feathers, or threads.

Leaves: Each 1- to 3-inch leaf is lanceolate to ovate and deeply lobed with gray pubescence on the underside. The foliage is aromatic when brushed. Its leaf arrangement is alternate.

Growing Needs: Garden mums prefer a sunny, fertile, well-drained location. Good drainage is important for winter survival. To maintain compactness, pinch plants when the foliage is 4 to 6 inches and once again after 4 to 6 inches additional growth. Discontinue pinching at the end of June to early July to allow for timely flower production. To ensure winter survival, do not cut the foliage back in the winter, plant mums in the spring or early summer for good root establishment, and mulch well.

Landscape Use:

Height:	6 inches to 3 feet
Habit:	Mounded
Spacing:	18 inches
Texture:	Medium

Plant the shorter garden mums in small groups (three to five) repeated throughout the sunny foreground for a show of color in the garden. Taller ones can be placed in the midborder or cut flower garden.

Design Ideas: Combine garden mums with asters, especially Purple Dome, sedums, and ornamental grasses. Or, try combining mums with poppies, yarrows, and daylilies to extend the flowering season. Hardy mum foliage will mask the fading foliage of early bulbs and poppies.

Cultivars: Numerous!

Problems: Aphids, leaf miners, powdery mildew.

Propagation: Division and spring to early summer tip cuttings.

Zones: 5-9

Echinacea purpurea

Purple Coneflower
Asteraceae (Sunflower Family)
○, ◑ Midborder, Background, Mid to late summer

Flowers: Each 3- to 4-inch flower head is daisy-like with rosy pink ray florets and a cone-shaped center. The older ray flowers droop back from the center. The disk flowers are bright orange then brown and form a mounded center.

Leaves: The 8- to 12-inch long petioled leaves of the basal rosette are ovate, toothed (denticulate), and have hispid pubescence (short stiff hairs). The stem leaves are alternate, sessile, rough, and lanceolate (narrower) with a less-toothed margin. The stems often have a dark color, striping, or flecking.

Growing Needs: This drought-tolerant plant is best sited in full sun but will tolerate light shade. The soil should be well drained. Deadhead flowers early to encourage new buds: later in the season, allow seed heads to form for some winter character. Taller plants may need peastaking or support from mounded perennials placed nearby.

Landscape Use:

Height:	3 feet
Habit:	Upright mounded
Spacing:	18 to 24 inches
Texture:	Medium bold

Purple coneflower fits well in the sunny midborder to background of the perennial bed. Plant it in a meadow or naturalized wildflower area.

Design Ideas: Purple coneflower, bronze fennel, white rose campion, and hardy geraniums all combine well together. It also combines well with blazing star and garden phlox. Use purple coneflower in front of ornamental grasses for a great textural contrast.

Cultivars:
White ones —

'Alba','White Swan', 'White Lustre'

Rosy pink or rosy purple ones —

'Bravado' is a large 6 inch in diameter rosy pink.

'Bright Star' is bright rosy pink.

'Crimson Star' is red flowering and grows 2 to 2½ feet tall.

'Magnus' is a deep rose with horizontal petals. 1998 Perennial Plant of the Year.

'Robert Bloom' is vigorous, long flowering, red purple, 2 to 3 feet tall.

Problems: None serious. (Leaf spot)

Propagation: Root cuttings, seed, and self sowing (not always true to type).

Zones: 4-8

Related Species:

E. pallida, Pale Coneflower, is a native prairie plant with light lavender pink drooping petals, 2 feet tall (June and July).

Echinops ritro

Globe Thistle
Asteraceae (Sunflower Family)
○   Background, Mid to late summer

Flowers: The 1½- to 2-inch flower head of globe thistle is an eye-catching, round, dense globose head that is steel blue in color. The stems are covered with a white pubescence. Attractive to bees!

Leaves: The leaves are produced in a rosette and alternate on the stem. Each 8-inch, dark green leaf is thistle-like and spiny with white pubescence underneath. The basal leaves are larger and petioled; the stem leaves are sessile. The leaf is pinnately dissected into toothed lanceolate lobes. Do not weed the emerging rosettes by mistake!

Growing Needs: Globe thistle needs a sunny, well-drained location for the best flowering. This plant can tolerate drought. Taller plants may need staking. Deadhead for a sparse reflowering; cut back to the rosette after flowering to prevent self sowing.

Landscape Use:

Height:	3 to 4 feet
Habit:	Upright mounded
Spacing:	2 feet
Texture:	Bold

Plant globe thistle in the sunny background, as a specimen or small group, or in the cut flower garden.

Design Ideas: Plant globe thistle with black-eyed Susan and Russian sage. It also looks great with bee balm (bee heaven!) and globe centaurea. Plant it with lamb's ear, 'Valerie Finnis' artemisia, baby's breath, or sea lavender to soften its texture. Try globe thistle and pink shrub roses together.

Cultivars:

'Taplow Blue' has large metallic blue flowers, 4 to 5 feet tall.

'Taplow Purple' has violet-blue flowers.
'Veitch's Blue' is dark blue in color and has gray-green leaves, 3 feet tall.

Problems: None serious.

Propagation: Seed, transplant self sown seedlings, careful spring division, root cuttings in spring.

Zones: 3-9

Related Species:

E. ruthenicus has bright blue, rounded flowers and chalky white stems. The leaves are more finely divided than *E. ritro*, 4 feet.

E. sphaerocephalus 'Arctic Glow' has a nearly white globular head, 2½ to 3 feet.

Eryngium amethystinum

Sea Holly
Apiaceae (Carrot Family)
○   Midborder, Mid to late summer

Flowers: Sea holly has a ½-inch, amethyst blue, ovoid-globose flower cluster. Each flower cluster is subtended by amethyst blue, thistle-like bracts. The stems are also silver blue.

Leaves: There are two types of sea holly leaves. The basal rosette leaves are 6 to 10 inches, ovate, and long petioled. The stem leaves are alternate, smaller, irregularly lobed (polymorphic), and clasping. Both types are spinose toothed, glossy, and medium green.

Growing Needs: This plant thrives easily in a full sun location with well-drained soil. The long root systems help it survive drought but makes it difficult to transplant.

Landscape Use:

Height:	2 feet
Habit:	Upright or two tiered
Spacing:	18 inches
Texture:	Medium bold

Place sea holly in the sunny midborder to background in small groups for accent. It also makes a good cut flower, both fresh and dried.

Design Ideas: Its unique texture combines well with delphinium, blazing star, and sea lavender. The silver artemisias and pink bee balm and Oriental lilies blend with sea holly for a soft pink, blue, and silver color combination with textural pizzazz.

Problems: None serious.

Propagation: Seed, careful transplanting of young self sown plants. Long tap root makes division very difficult.

Zones: 3-8

Related Species:

E. bourgatii has larger silvery blue flowers, blue stems, and showy white-veined, dissected spiny leaves, 18 inches to 2 feet.

E. alpinum has showy blue, thistle-like flowers, longer petioled, finer-textured leaves, and green stems. 'Blue Star' has deep blue large flowers, 2 feet.

E. giganteum, Miss Wilmott's Ghost, is 4 to 5 feet with large pale blue flowers.

E. planum is taller with green stems.

'Blaukappe' is a shorter cultivar at 24 to 30 inches with bright blue flowers.

E. yuccifolium, Rattlesnake Master, is 3 to 5 feet with narrow yucca-like foliage and clusters of white dome-shaped flowers with spiny bracts; native.

Eupatorium maculatum

Joe-Pye Weed, Boneset
Asteraceae (Sunflower Family)
○ Background, Late summer to fall

Flowers: Joe-pye weed has 10- to 12-inch clustered (cymose) panicles of pale rose to light purple.

Leaves: The foliage of boneset is whorled (three to four) around the stem, ovate to lanceolate, and 8 to 10 inches long, decreasing in size up the stem. Each leaf is coarsely serrated. The stems are speckled with purple. Rub the foliage for a vanilla scent.

Growing Needs: Joe-pye weed does well in full sun with moist, well-drained soils. Like asters and mums, prune it back once in the spring when it is 1 foot high.

Landscape Use:

Height:	4 to 6 feet
Habit:	Upright
Spacing:	2½ feet
Texture:	Bold

This bold, tall plant can easily be placed in the sunny background of the informal or formal garden. It also naturalizes nicely in the meadow or wildflower garden.

Design Ideas: Plant Joe-pye weed with purple coneflowers, garden phlox, and ornamental grasses. Boltonia, Russian sage, and asters look especially effective when placed in front of Joe pye-weed.

Cultivars:

'Atropurpureum' has large, dark purple flowers with purple-flecked stems and leaf veins.
'Gateway' is 4 to 5 feet with large rounded, rosy pink flower clusters and reddish stems.

Problems: None serious.

Propagation: Division, cuttings, seed.

Zones: 3-8

Related Species:

E. purpureum is 5 to 7 feet with smaller purple flowering clusters. The stems are green but may have some purple mottling at the nodes.

E. coelestinum, Hardy Ageratum or Mistflower, is now *Conoclinium coelestinum*. It is a vigorous, long-flowering, 2- to 3-foot plant with ageratum blue fluffy flower clusters. 'Alba' has white flowers. 'Cori' is 2 feet and long flowering.

E. rugosum 'Chocolate' is now *Ageratina altissima* 'Chocolate'. It is distinctive for its chocolate brown foliage, purple stem, 4 to 5 foot height, and contrasting white corymbs.

Helenium autumnale

Helen's Flower, Sneezeweed
Asteraceae (Sunflower Family)
○　Midborder, Background, Late summer to fall

Flowers: These bright yellow, orange, or bronze daisy-like flower heads are solitary or in corymbs and 1½ to 2 inches wide with a prominent mounded center. The three-toothed ray florets fling back or droop down from the center.

Leaves: The 2- to 6-inch leaves are alternate, ovate to lanceolate, and have toothed margins. The glossy, nearly glabrous leaves are decurrent (extend down the stem) and make the stems appear winged. (Great ID feature!) The lower leaves are often deciduous.

Growing Needs: A sunny, moist, well-drained site is required for Helen's flower. The plants will languish or die in hot, dry sites. Mulch and water as needed. Pinch back new shoots in late spring for a more branched plant. Stake taller plants.

Landscape Use:

Height:	2 to 6 feet
Habit:	Upright mounded
Spacing:	2 to 2½ feet
Texture:	Medium

Use this plant massed in the sunny background border or in the cut flower garden.

Design Ideas: Bronze-red Helen's flowers look great near sunflower heliopsis, *Rudbeckia nitida* 'Autumn Sun', and Silver Feather maiden grass (bronze fall color) with blue asters or veronicas in front. Russian sage or blue beard (*Caryopteris*) contrast nicely with it.

Cultivars:

'Brilliant' is 3 foot, well branched, floriferous, bronze red.

'Bruno' is 4 foot, crimson flowered.

'Butterpat' is yellow and 3 feet tall. 'Brillant' has chestnut colored flowers. 'Moerheim Beauty' has bronze-red flowers, which fade to a burnt orange color. It is low maintenance and 3 feet.

'Redgold Hybrids' has a mixture of red, bronze, orange, gold, and yellow flowers with some bicolors, 2 to 4 feet.

'Riverton Beauty' has yellow flowers with a dark reddish brown center, 4 feet.

'Rubinzwerg' is dark red and 3 feet tall.

'Wyndley' is a compact 2½ to 3 feet and has striking coppery golden flowers.

Problems:　Powdery mildew (hot, dry)

Propagation:　Division. (Seed—not cultivars)

Zones:　3-8

Heliopsis helianthoides

Sunflower Heliopsis, False Sunflower
Asteraceae (Sunflower Family)

○ Midborder, Background, Mid to late summer to fall

Flowers: This daisy-like flower head is 2 to 3 inches across and deep yellow. Disk flowers are slightly mounded. Each solitary flower can be single, semi-double, or double.

Leaves: The foliage is opposite (occasionally whorled), ovate to lanceolate with toothed margins, and petioled. Each 3- to 5-inch leaf is thin, dark green with an obvious midvein and gritty (scabrous) pubescence (stem is glabrous).

Growing Needs: This easy-to-grow plant will thrive in a sunny, well-drained, moderately fertile site. Sunflower heliopsis is tolerant of drought and poor soils. Tall cultivars may require staking. Remove faded flowers to prevent excessive self sowing.

Landscape Use:

> Height: 3 to 5 feet
> Habit: Mounded
> Spacing: 18 to 24 inches
> Texture: Medium bold

Sunflower heliopsis will fit nicely into the sunny midborder or background of the perennial border or cut flower garden. Use it fresh or dried.

Design Ideas: This long-flowering plant combines well in a hot, hot, hot color scheme. Plant with red lilies, orange butterfly flower, yellow tickseed, red hot poker, and Maltese cross.

Cultivars:

'Ballerina' is yellow, semi-double, 3 feet tall.
'Gold Greenheart' is a fully double yellow flower with green centers, 3 feet tall.

'Gold Plume' has double yellow flowers, is 3½ feet, and floriferous.
'Karat' is 4 feet, single yellow, long flowering.
'Mars' is yellow-orange, 4 to 5 feet.
'Summer Sun' has bright yellow semi-double flowers, 3 feet.

Problems: None serious.

Propagation: Division, seed (self sows), and cuttings.

Zones: 3-9

Related Species:

H. helianthoides var. *scabra* has coarsely hairy, scabrous leaves and stems.

Hosta sp.

Hosta, Plantain Lily, Funkia
Liliaceae (Lily Family)
◐ (○)   Foreground to background, Mid to late summer

Flowers: Hosta produces one-sided racemes up to 3 feet long. Each 1- to 4-inch, trumpet-shaped flower is either white or shades of lavender. Some of the flowers are very fragrant.

Leaves: This plant is grown for its beautiful 6- to 12-inch, cordate to lanceolate leaves. Each leaf has a long furrowed petiole and deep venation. The leaves come in solid, variegated, and mottled colors including green, blue-green, blue, chartreuse, yellow, gold, and bicolors.

Growing Needs: This is an excellent plant for partial shade (or sun, depending upon the cultivar). When sited in full sun, the soil should be moist. Hosta prefers a fertile, deep, well-drained soil. Though rarely needing it, hosta can be divided in early spring at the pointed bud stage. Hostas look better with age.

Landscape Use:

Height:	1 to 3 feet
Habit:	Mounded
	Two tiered (in flower)
Spacing:	3 feet
Texture:	Medium to bold

Plantain lily is an exceptional addition to the shady garden as a specimen or massed planting.

Design Ideas: Plant plantain lily with ferns, Bethlehem sage, purple-leaved coral bells, and European wild ginger. It is great for under the canopy of trees.

Specific Species/Cultivars (just a few):

H. fortunei has bold green/blue-green leaves and pale lavender flowers, 2 feet tall.

'Aureo-marginata', 'Hyacinthina'

H. plantaginea, Fragrant Plantain Lily, has large green to yellow green leaves, large white flowers, and grows 2 feet tall. 'Aphrodite' — double white flowers.

H. sieboldiana has very large, bold, gray-green or blue-green leaves, lavender flowers, and grows 3 feet tall. 'Elegans' — blue-leaves.

'Frances Williams' (a sport of *H. s.*) — blue leaves and irregular golden edges, white flowers. Truly beautiful.

'Golden Sunburst' (a sport of Frances Williams) — gold leaves.

'Great Expectations' (a sport of *H. s.* 'Elegans') — irregular coloration of blue and gray-green with a golden center, 2 to 3 feet.

H. tokudama has bold puckered foliage, white flowers, and grows 2 feet.

var. *aureo-nebulosa* — yellow center, blue-edged leaves.

'Love Pat' — puckered, cupped, dark-blue foliage, somewhat slug resistant.

H. undulata, Wavy Leaf Hosta, has wavy variegated foliage, lavender flowers, 1 to 2 feet. var. *albo-marginata* has a creamy white margin.

var. *univittata* has a green and white striped effect, both colors extend down the wavy petiole, 2 to 3 feet.

Other Fine Cultivars:

'Big Daddy' — 3 feet, large, blue-green puckered Hosta, slug resistant.

'Blue Angel' — largest of the blue Hostas, space it 4 to 5 feet, white flowers.

'Francee' — nice, white-edged green foliage, lavender flowers, 2 feet.

'Ginko Craig' — small, 10-inch plant for edging, green leaves, white margins.

'Gold Standard' — golden green leaves with dark-green margins, 2 feet, showy.

'Golden Tiara' — compact, medium green leaves with gold margins, 1 foot.

'Great Expectation' — a sport of *H*

'Halcyon' — chalky blue, puckered foliage, lavender flowers, 18 inches.

'Krossa Regal' — blue leaves, lavender flowers, vase-shaped habit, 2½ feet.

'Little Aurora' — small, gold leaves, 8 inches.

'Patriot' — sport of France with green foliage and wide (1 inch) white margins, 1997 Hosta of the Year.

'Royal Standard' — solid light-green leaves, fragrant white flowers, 2 to 2½ feet.

'So Sweet' — green foliage, white margins, fragrant, near-white flowers.'Sum and Substance' — yellow-green, glossy foliage, 3 feet.

'Wide Brim' — blue-green foliage with wide, irregular, creamy white margins, 18 to 24 inches.

Problems: Slugs, crown rot, leaf spot.

Propagation: Early spring division and seed (patience required—3 to 4 years to flower).

Zones: 3-8

Hylotelephium

(Sedum) species and hybrids

Showy Sedum or Stonecrop
Live-Forever
Crassulaceae (Stonecrop Family)
○, ◑ Foreground, Late summer to fall

Flowers: Stonecrop flowers are produced in 3- to 6-inch, dense, flat-topped to slightly rounded cymes. Each flower is ½ inch, star shaped, and pink, red, or white. Even the young flower clusters of Autumn Joy showy sedum are striking, resembling broccoli. The flowers turn bronze, then brown, and persist throughout the winter.

Leaves: The succulent leaves are 1 to 3 inches, sessile, ovate with margins that are either toothed (*H. telephium*) or scalloped (*H. spectabile*). The leaves of *H. spectabile* are opposite or in whorls; the leaves of *H. telephium* are alternate. The nearly emerging foliage is very attractive, looking whorled; the summer foliage appears slightly cupped.

Growing Needs: This drought-tolerant plant thrives in well-drained soils with full sun to light shade. Do not trim back the faded flowers since these stems provide winter character and winter protection for the crown; leave the stems until early spring and then remove.

Landscape Use:

Height:	18 to 24 inches
Habit:	Mounded
Spacing:	18 inches
Texture:	Medium bold

This tough plant fits nicely in the foreground in small groupings, as a specimen planting, or in the rock garden. The flowers are good as a fresh or dried cut flower. Butterflies love this plant as a late nectar source.

Design Ideas: Mix showy sedum with other low-growing sedums and blue balloon flowers. A favorite combo is with blue plumbago and woolly thyme (edging), showy sedum and summer snow obedient plant (foreground), Monch aster and Russian sage (midborder), and white boltonia and maiden grass — *Miscanthus sinensis gracilimus* (background).

Problems: None serious.

Propagation: Summer stem cuttings (new plants rooted for me in the fall from flowering stems jokingly placed in the soil nearby and forgotten until the next spring) and division (rarely needed).

Zones: 3-8

Cultivars of Hylotelephium spectabile:

Late summer to fall

Height ranges from 15 to 24 inches.

'Album' white.

'Brilliant' deep pink.

'Carmen' rose pink.

'Iceberg' white.

'Indian Chief' deep pink-red.

'Meteor' deep carmine-red.

'Snowqueen' white.

'Stardust' ivory white flowers with light green leaves.

'Variegatum' pink flowers with green and creamy white foliage.

Cultivars of Hylotelephium telephium:

Late summer to fall

Height ranges from 18 to 24 inches.

ssp. *maximum* 'Atropurpureum' deep rose red flowers with burgundy-colored foliage, somewhat floppy.

'Honeysong' rose pink flowers with dark reddish-purple leaves, 18 inches.

'Munstead Red' deep red flowers with bronzy purple foliage.

Other Hybrids:

Parentage noted when known.

Late summer to fall unless noted.

'Autumn Joy' ('Herbstfreude')

(*H. telephium* x *H. spectabile*) deep rosy pink flowers, turning salmon to bronze to brown for winter character. Sturdy mounded habit, 18 to 24 inches.

'Bertram Anderson' red flowers with purple leaves, best suited to full sun for the best foliage color, 6 to 8 inches.

'Frosty Morn' pale pink to white flowers with light green foliage and wide white margins, 12 inches.

'Matrona' pale pink flowers with dark gray foliage edged in pink, red stems, 2 to 3 feet.

'Mohrchen' early fall flowering pale red flowers with burgundy leaves, 2 feet.

'Ruby Glow' ('Rosy Glow') (*H. cauticolom* x *H. telephium*) deep ruby red with gray-green tinged red. Mid to late summer, 8 inches.

'Vera Jameson' (*H. telephium* ssp. *maximum* 'Atotpurpureum' x 'Ruby Glow') pale dusty pink flowers with purple leaves, 10 to 12 inches.

Related Species:

Hylotelephium sieboldii has pink flowers in fall, gray-green rounded foliage with trailing stems, 6 to 8 inches.

H. sieboldii 'Medio-variegatum' has a creamy yellow band in the center of the blue-gray leaf, 6 to 8 inches.

*K*niphofia hybrids

Red Hot Poker, Torch Lily, Tritoma
Liliaceae (Lily Family)
○ Midborder, Background, Mid to late summer

Flowers: Torch lily makes a great show in the garden with its 6- to 8-inch tapering racemes of dense, drooping tubular flowers. Each flower is 1 to 2 inches long and can be cream, yellow, orange, red, and two-toned. Hummingbirds adore these tubular flowers.

Leaves: The gray-green linear (1-inch wide) leaves are fleshy, grass-like, and keeled (V-shaped) with rough margins. The long (2- to 3-foot) leaves are borne in a basal rosette and are arching or gracefully bending by midsummer. Remove in early spring before new growth emerges.

Growing Needs: This full-sun plant must have a well-drained soil. Red hot poker is very low maintenance and rarely needs dividing except to propagate it.

Landscape Use:

Height:	2 to 4 feet
Habit:	Two tiered
Spacing:	2 feet
Texture:	Bold

Torch lily makes an excellent accent grouping in the sunny midborder of the garden. Place it in front of shrubs or in the cut flower garden.

Design Ideas: Combine torch lily with yellow yarrows, garden phlox, and boltonia with catmint as edging. Yellow torch lily looks great when planted with purple coneflowers or liatris.

Cultivars: Numerous!

'Ada' has gold spikes on 3- to 3½-feet stems.
'Atlanta' has early bright yellow flowers, 3 feet.
'Earliest of All' has early coral red flowers and is 2½ feet.

'Little Maid' is 2 feet with white flowers, tipped pale yellow.
'Pfitzeri' has dark orange flowers, 30 inches.
'Springtime' is a bicolor of coral red (top) and pale yellow, 3½ feet.

Problems: None serious.

Propagation: Seed and spring division.

Zones: 5-9

Related Species:

K. uvaria, a parent of the red hot poker hybrids, is red (top) and yellow, and 4 feet tall.

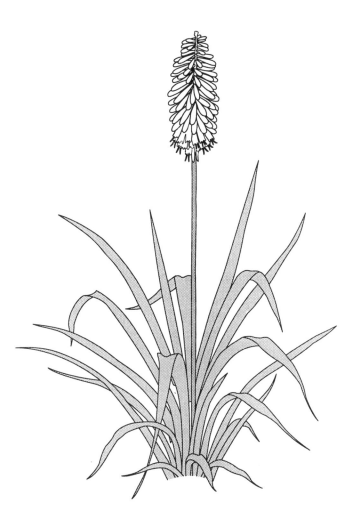

Liatris spicata

Blazing Star, Liatris, Gayfeather
Asteraceae (Sunflower Family)
○, ◑ Midborder, Mid to late summer

Flowers: The fuzzy, rosy purple or white flowers are densely arranged on a 6- to 12-inch spike. Unlike most other spike flowers, these flower from the top down, not from the bottom up.

Leaves: The narrow lanceolate leaves are 1 to 2 inches along the stem and up to 10 inches at the base. The leaves are alternate to whorled on the stem, resembling narrow lily foliage. The leaf surface is glabrous or rarely hirsute (short, stiff hairs) and punctate (glandular pits).

Growing Needs: This native plant likes a full sun to part shade site. A well-drained soil is a must. Trim back faded flowers to avoid self sowing (or to collect the seeds).

Landscape Use:

Height:	2 to 4 feet
Habit:	Upright
Spacing:	12 to 18 inches
Texture:	Medium fine

Use blazing star as a accent grouping in the sunny midborder to back-ground of the garden. It also makes an excellent long-lasting cut flower and dries well.

Design Ideas: Blazing star looks great with purple coneflower and *Coreopsis* 'Moonbeam'. It also combines well with ornamental grasses, silver artemisia, baby's breath, or white campion.

Cultivars:

'Alba' is a white form.
'Kobold' is a compact 18 inches to 2 feet and produces many rose-lavender flowers.
'Floristan Violet' has rosy violet flowers, 3 to 3½ feet.
'Floristan White' has white flowers, 3 feet.

'Silver Tip' has lavender flowers.

Problems: None serious.

Propagation: Seed and division.

Zones: 3-9

Related Species:

L. microcephala is a 1-inch miniature version with thin, rosy lavender spikes.

L. pycnostachya, Kansas Gayfeather, has showy, 4-feet, lavender spikes.

L. scariosa 'September Glory' is a later flowering lavender spiky plant for the garden, 3 to 4 feet. 'White Spires' is the white version.

Limonium latifolium

Sea Lavender
Plumbaginaceae (Plumbago Family)
○   Foreground, Mid to late summer

Flowers: The flowers of sea lavender are borne in a 1- to 2-foot, multi-branched, rounded panicle. Each individual ¼-inch flower is surrounded by a white calyx with a lavender corolla, fading to white. The overall appearance is quite airy and softening.

Leaves: The leaves arise from a basal rosette. Each leaf is oblong, elliptic or spatulate (spoon-shaped), leathery, and up to 10 inches long.

Growing Needs: This full-sun plant does well in light, sandy-loam soils. It is also considered salt and pollution tolerant. Sea lavender is a low-maintenance plant and rarely needs division except to propagate it.

Landscape Use:

Height:	1 to 2 feet
Habit:	Two tiered
Spacing:	18 inches
Texture:	Fine

Site sea lavender in the sunny foreground. It makes an excellent softening plant when used with bold or stiffly vertical plants. Find room for sea lavender in the cut flower garden: it is an outstanding cut and dried flower and a favorite florists' filler flower.

Design Ideas: Sea lavender is attractive when combined in front of bold garden phlox and vertical blazing star with medium-textured *Dianthus* as edging. Place sea lavender with poppies to soften it and hide its lack of summer foliage.

Cultivars:

'Violetta' has a violet blue flower.
'Collier's Pink' has a pink flower.

Problems: None serious.

Propagation: Seed and spring division.

Zones: 3-9

Related Species:

Formerly *Limonium tatarica or dumosum*, *Goniolimon tataricum* is the correct name for a beloved fresh and dried cut flower—German statice. It is covered with lavender blushed, starry white flowers in early summer.

Limonium gmelinii, Siberian Statice, produces darker purple-blue flowers than *L. latifolium*. 18 inches high.

Lysimachia clethroides

Gooseneck Loosestrife
Primulaceae (Primrose Family)
◯, ◐ Midborder, Background, Mid-summer

Flowers: Gooseneck loosestrife has white, ½-inch, star-shaped flowers on very distinctive and graceful arching racemes, 3 to 6 inches long. This gooseneck appearance is one-of-a-kind among flowers. It is a lovely long-lasting cut flower.

Leaves: The alternate leaves are ovate to lanceolate, taper at both ends, and are 2 to 5 inches long. The pubescent leaves may emerge with a red edge. In the fall, the foliage turns a lovely bronze-yellow.

Growing Needs: This plant is best sited in a full-sun, well-drained spot. It is an assertive plant, but is easy to manage by pulling the excess stems or dividing often. In a moist, partially shady location, it will become a very aggressive spreading perennial. Plant in a confined spot for lower maintenance.

Landscape Use:

Height:	2 to 3 feet
Habit:	Mounded
Spacing:	2 to 3 feet
Texture:	Medium

Gooseneck loosestrife fits well into the sunny midborder to background of the perennial border. It makes a good accent planting. The curving flowers seem to blend with other shapes well.

Design Ideas: Plant gooseneck loosestrife next to globe centaurea, fernleaf yarrow, and red hot poker with perennial blue salvia and Missouri sundrops in front of it. Siberian iris foliage, Russian sage, lamb's ear as edging also blend with the texture and shape of gooseneck loosestrife.

Problems: None serious.

Propagation: Spring or fall division and seed.

Zones: 3-9

Related Species:

L. ciliata 'Autropurpureum', Fringed Loosestrife, is a 3-foot plant with purple leaves (with ciliate or fringed margins) and canary yellow flowers. A spreading plant for sun or shade.

L. punctata 'Alexander' has variegated foliage (green with creamy margins), low-key yellow flowers, and is less invasive. 18 to 24 inches.

Lythrum virgatum

Purple Loosestrife
Lythraceae (Loosestrife Family)
○ Midborder, Background, Mid to late summer

Flowers: Purple loosestrife produces linear 12-inch spikes of reddish-purple or magenta flowers. Individual florets are nearly 1 inch across.

Leaves: Each 3- to 6-inch leaf is lanceolate, sessile and non-clasping, and opposite or whorled around the stem.

Growing Needs: Purple loosestrife should be in a full-sun, well-drained site.

Landscape Use:

Height: 2 to 4 feet
Habit: Upright to mounded
Spacing: 2 feet
Texture: Medium

Plant purple loosestrife massed in the sunny midborder or background for showy color in the summer.

Design Ideas: This plant combines well with Powis Castle artemisia, late yellow daylilies, and purple coneflower. It is also beautiful when interplanted with Russian sage.

Cultivars:

'Dropmore Purple' is 3 feet with purple flowers.
'Morden Gleam' is deep rose-pink.
'Morden Pink' is pink to magenta.
'Morden Rose' is rose-red.
'Rose Queen' is 2 to 3 feet tall and pink.
'The Rocket' has vibrant rose-pink flowers.

Problems: None serious.

Propagation: Division and cuttings (no seed).

Zones: 3-10

Related Species:

Lythrum salicaria is a taller, less refined plant with wider, clasping leaves. It has become a serious invasive problem in northern states' wetlands areas and is illegal to sell in many states, including the cultivars, such as 'Happy', 'Robert', 'Firecandle', 'Brightness', 'The Beacon', 'Lady Sackville', and 'Atropurpureum'.

FYI: The flowers of this *L. virgatum* generally do not produce seed. Some studies have shown that *L. virgatum* may cross with the invasive *L. salicaria* and produce seeds and fertile offspring. Many nurseries are no longer selling any *Lythrum* species or cultivar because of the controversy.

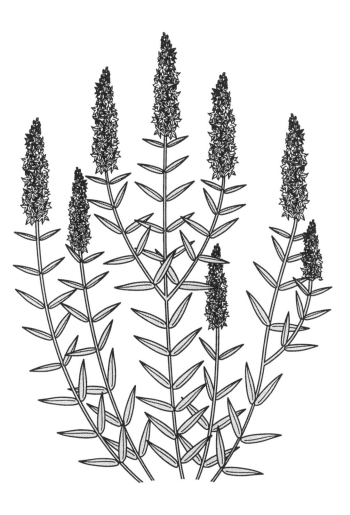

Perovskia atriplicifolia

Russian Sage
Lamiaceae (Mint Family)
○, ◑ Midborder, Background, Mid to late summer to fall

Flowers: Each ½-inch flower is two lipped and lavender to blue in color. After the petals fall off, the calyx is still showy, densely pubescent, and lavender-blue. The flowers are borne in profusion on 12- to 24-inch spiky upright panicles.

Leaves: The 1- to 2-inch leaves are opposite, grayish-green above and whitish below. The species has finely dissected; the cultivars vary from entire to very finely dissected foliage. The leaves and the white stem are aromatic. The surface of the leaf feels "sticky" to the touch.

Growing Needs: Russian sage prefers full sun and well-drained, light soils, even sandy or gravelly ones. It will tolerate partial shade but will have a floppy habit, requiring staking. Leave the attractive white cloud-like clumps to enjoy during the winter. Cut back this semi-woody plant to 4 to 6 inches in early spring before new growth starts.

Landscape Use:

Height:	3 to 4 feet
Habit:	Upright
Spacing:	18 to 24 inches
Texture:	Medium fine to fine

Plant Russian sage in the sunny background or as the center of an island bed. It provides a lovely vertical accent in the summer and winter garden. Russian sage seems to blend with everything because of its texture and coloration. When used as a fresh or dried cut flower, this plant sheds and is very messy.

Design Ideas: Russian sage combines well with veronicas, lilies and, black-eyed Susan. A fa-vorite combination for a knock-out fall garden is blue plumbago, dusty miller, and annual Victoria blue salvia (edging), Autumn Joy sedum (foreground), Benary hybrid asters and Vivid obedient plant (midborder), and Russian sage (background). The 1995 Perennial Plant of the Year combines well with so many shrubs, perennials, annuals, and herbs. It is a must in the garden!

Cultivars:

'Blue Mist' has light blue flowers and nearly eantire leaves.

'Blue Spire' has larger lavender-blue flowers and finely dissected leaves.

'Filigran' is more finely dissected than the others.

'Longin' has an erect habit and less dissected leaves.

Problems: None serious.

Propagation: Cuttings and transplanting of small offshoots.

Zones: 5-9

Related Species:

P. abrotanoides has similar flowers but more deeply dissected leaves and a wider, more branched habit. It is possibly a parent of some of the *P. atriplicifolia* cultivars.

P. scrophulariaefolia has a 3- to 4-foot, semi-woody, shrubby habit with darker blue to purple flowers and ovate leaves.

Phlox paniculata

Garden Phlox
Polemoniaceae (Phlox Family)
○, ◑ Midborder, Background, Mid to late summer

Flowers: Garden phlox produces large, rounded, 6- to 10-inch panicles. Each five-lobed, 1-inch flower ranges in color from white, pink, red, purple, and bicolors, some with darker eyes. The buds form attractive pointed clusters.

Leaves: The 3- to 4-inch leaves are thin, pointed, ovate to lanceolate, and opposite with obvious venation. The margins are ciliate and minutely toothed. Red or purple flowering varieties often have reddish leaves or venation. During drought or other stress, phlox leaves will turn yellow and drop off.

Growing Needs: The best flowering and growth for phlox is in partial shade or sun in moist, fertile, well-drained soils. This plant will show signs of stress under drought conditions. Stake larger plants and remove faded flowers. On mature plants, select six to eight strong stems and thin out weaker stems to produce larger flowers and to promote air circulation, which helps to prevent mildew. Avoid watering the leaves to avoid mildew problems. Provide generous space between plants for good air circulation. Select mildew resistant varieties.

Landscape Use:

Height:	2 to 4 feet
Habit:	Mounded to upright mounded
Spacing:	18 inches
Texture:	Medium to bold

Plant garden phlox in masses in the sunny midborder to background.

Design Ideas: Plant garden phlox with *Veronica* 'Sunny Border Blue' or 'Icicle'. It also combines well with liatris, gooseneck loosestrife, purple coneflower, and catmint (for edging). Miss Lingard phlox is striking with perennial blue salvia, red bee balm, and red coral bells.

Cultivars:

White —

'David' highly mildew resistant, 3 to 4 feet.

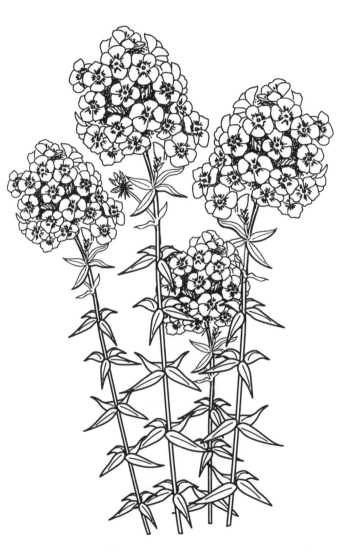

'Mt. Fujiyama' is a pure white, 3 feet.

Pink—

'Bright Eyes' with red eye, 2 feet.

'Eva Cullom' with red eye and mildew resistant, 2 to 3 feet.

'Fairest One' light salmon-pink, 3 feet.

'Pinafore Pink' bright pink, 8 to 12 inches.

'Prime Minister' whitish pink with a dark pinkish red eye, 3 feet.

Dark Pink/Red—

'Starfire' showy dark rosy red with deep red foliage, 2 to 3 feet.

'Sandra' dark rosy red, 2 feet.

Orange Red—

'Orange Perfection' orange-red flowers, 3 feet.

Lavender—

'Blue Boy' lavender blue, 3 feet.

'Franz Schubert' lavender lilac, 3 feet.

'The King' dark purple, 3 feet.

Variegated Leaves—

'Darwin's Joyce' green foliage with creamy yellow edge, pink flower with dark pink eye, 2 feet.

'Norah Leigh' gray-green foliage with white edge, pinkish lavender flower with pink eye, 2 to 3 feet.

Problems: Powdery mildew, root rot, and red spider mites.

Propagation: Cuttings and division.

Zones: 4-8

Related Mildew Resistant Species:

Phlox maculata, Meadow Phlox, has compact, conical flowers and darker glossy green leaves, which are very mildew resistant.

'Alpha' rosy pink, 2½ feet.

'Delta' white with a pink eye, 3 feet.

'Miss Lingard' white, 2 to 3 feet (may be listed a *P. carolina*).

'Natascha' lavender pink and white bicolor, 2 to 2½ feet.

'Omega' white with rose eye, 3 feet.

'Rosalinde' dark pink, 3 feet.

P. carolina, Thick-leaf phlox, has a glossy thick leaf with indistinct veins. The thick leaf makes it highly mildew resistant. Thick-leaf phlox is earlier flowering than *P. paniculata*. The flower colors are pink to purple.

'Gloriosa' is salmon-pink, 3 to 3½ feet.

Physostegia virginiana cultivars

Obedient Plant, False Dragonhead
Lamiaceae (Mint Family)
○, ◑   Midborder, Background, Late summer to fall

Flowers: The 1-inch flowers are borne in four vertical rows on thin, 6- to 8-inch racemes. Additional smaller racemes are borne in the leaf axils and prolong the flowering. The colors range is white to shades of pink.

Leaves: Each medium-green, glossy leaf is opposite and sessile, 3 to 5 inches long, and lanceolate with sharply serrate margins. A great identification feature is the obvious square stem.

Growing Needs: This vigorous plant does well in full sun to partial shade. Any well-drained soil will do. To keep it where you want it, consider planting it in a confined spot or placing it in a buried pot with the bottom removed. The leaves easily cover the pot edge and you have no worries. 'Summer Snow' and 'Variegata' are less aggressive. Otherwise, check the plant yearly and divide it to keep it the size that you want. Obedient plant is easy to pull out. It is worth the minimal effort.

Landscape Use:

> Height: 2 to 3½ feet
> Habit: Upright
> Spacing: 2 feet
> Texture: Medium

False dragonhead looks best in small groups in the sunny midborder of the garden or in the cut flower garden. It has a cottage garden look and works well naturalized in a meadow or wood-land edge area. Plant it on slopes or in poor soil.

Design Ideas: Place this plant in the border with tough plants that will not be pushed around, like Autumn Joy sedum, asters, and *Heliopsis*. Try it with artemisias and *Centranthus*.

Cultivars of *P. v. ssp. praemorsa:*

'Alba' has white flowers, 2 feet.

'Bouquet Rose' is lavender pink, 3 feet.

'Pink Bouquet' has rose pink flowers.

'Summer Snow' has a pure white flower and is less vigorous, 2½ feet.

'Variegata' has later flowering pink flowers with green and gray-green leaves edged in creamy white, 2 to 3 feet.

'Vivid' has deep rose pink flowers and is very late flowering, 18 to 24 inches.

Problems: None serious.

Propagation: Division and cuttings.

Zones: 3-9

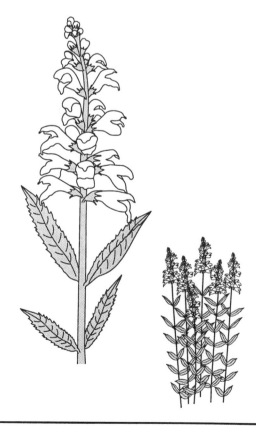

Platycodon grandiflorus

Balloon Flower
Campanulaceae (Bellflower Family)
○, ◑ Midborder, Mid to late summer

Flowers: The buds of this flower look like a small hot air balloon. They pop when squeezed (a favorite activity of my 7-year-old son). Each 2-inch blue flower is solitary (or in small flowered corymbs) and bell-shaped with darker or contrasting venation. Cultivars can be pink or white.

Leaves: The medium-green foliage is alternate, 1 to 3 inches long, lanceolate to elliptic-lanceolate, and sharp toothed. The glabrous foliage emerges late in spring; take care not to harm the crown.

Growing Needs: A moderately fertile, neutral to slightly acidic, well-drained soil in a sunny to partially shaded location is suitable for balloon flower. Taller flowers may need peastaking or placed next to sturdy plants.

Landscape Use:

Height:	2 to 3 feet
Habit:	Vase shaped
Spacing:	18 inches
Texture:	Medium

Plant balloon flower massed in the sunny midborder. This plant also makes a welcome addition to the cut flower garden.

Design Ideas: Balloon flowers look great when planted near Russian sage, fall anemones, anise hyssop, and lamb's ear. I like it in my blue and green garden(with yellow accents) with catmint, lady's mantle foliage, peach-leaved bellflower, and the silver foliage of moonshine yarrow.

Cultivars:

'Album' has white flowers, 2 to 2½ feet.
'Apoyama' is lavender blue, 9 inches.
'Double Blue' is 18 to 24 inches and has double blue flowers.
'Komachi' have large inflated blue balloon-like buds, which never open.
'Lavender' has lavender flowers, 18 inches.
'Mother of Pearl' has large pink flowers, 3 feet.
'Roseus' is a pink flowering plant.
'Sentimental Blue' is a mounded 18 inches.
'Shell Pink' has light pink flowers, 2 feet.
var. *mariesii* is a compact 18 inches with deep blue flowers.

Problems: None serious.

Propagation: Seed (self sow) or spring division.

Zones: 3-9

Stachys byzantina

Lambs Ear, Betony
Lamiaceae (Mint Family)
○, ◑, ●   Edging, Foreground, Ground cover,
Mid-summer

Flowers: The asymmetrical, ½-inch flowers are
lavender pink and whorled on woolly white flowering spikes. The flowers are attractive to bees and are
often removed because of the bees and to enjoy the
foliage rosette. The 8- to 12-inch spikes are excellent for fresh and dried cut flowers.

Leaves: Each 4- to 6-inch leaf is thick and covered with white tomentose pubescence. It is so soft
that it feels like fur. The leaves are oblong, elliptic
(narrowed at each end), and have slightly crenate
(scalloped) margins and white woolly petioles.

Growing Needs: Lamb's ear does best in a
sunny, well-drained area. It will also grow in partial
shade, even dense shade, though flowering is re-
duced. Lamb's ear does not perform well in hot,
humid or poorly drained conditions because the
leaves rot. Some gardeners feel that the flowers de-
tract from the overall foliage appearance and remove them.

Landscape Use:

Height:	6 to 10 inches
	(18 inches in flower)
Habit:	Low mounded,
	two tiered in flower
Spacing:	18 to 24 inches
Texture:	Medium

Lamb's ear makes a good sunny edging or accent
plant. Use it as a limited ground cover in conjunc-
tion with darker foliaged plants.

Design Ideas: This attractive foliage plant
looks great next to creeping phlox with hardy geraniums (Johnson's Blue or Mayflower), iris foliage,
and sea holly (*Eryngium*) behind it. Contrast it with
burgundy-leaved coral bells or basils. Plant it with
Caryopteris × *clandonensis* 'Longwood Blue'.

Cultivars:

'Big Ears' has very large leaves and very few flowers, 10 inches.
'Helene von Stein' has very large, less woolly leaves
and very few flowers, 8 inches.
'Silver Carpet' is non-flowering.
'Primrose Heron' has yellow spring foliage that
turns gray green in the summer.

Problems: None serious.

Propagation: Spring division and seed.

Zones: 4-8

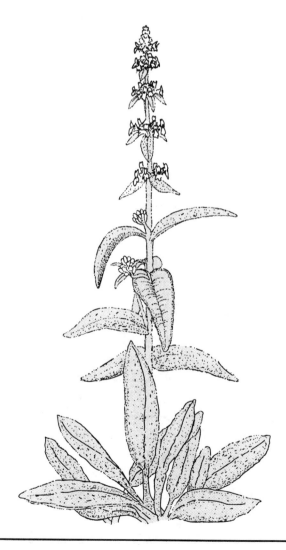

Veronica spicata

Veronica, Spike Speedwell
Scrophulariaceae (Snapdragon Family)
○, ◑ Midborder, Mid to late summer

Flowers: The 6- to 8-inch long, tapering racemes are covered with ¼-inch flowers up to the tip. Smaller racemes are formed in the leaf axils to prolong the flowering. The flowers come in shades of blue, rose-pink, or white. Removing spent flowers will lengthen the flowering season.

Leaves: The foliage is opposite, ovate to lanceolate, with crenate margins. The 2- to 3-inch leaves are pubescent (or may be glabrous and glossy depending upon the cultivar or hybrid).

Growing Needs: Plant veronica in a sunny site with well-drained soil. It will tolerate partial shade but does not flower its best.

Landscape Use:

Height:	18 to 24 inches
Habit:	Upright mounded
Spacing:	18 inches
Texture:	Medium

Place veronica in the sunny foreground to midborder of the garden or cut flower border for long flowering in the summer to early fall. The blue color is welcome in the late summer garden.

Design Ideas: The spiky flowers look great with round flowers like coreopsis, golden chamomile (*Anthemis*), or Shasta daisies, as well as with unique shapes like yarrows, garden phlox, or daylilies.

Cultivars:

'Alba' is the white form.
'Blue Fox' is lavender blue, 12 to 15 inches.
'Blue Peter' is deep blue, 18 to 24 inches.
'Caerulea' has sky blue flowers.
'Erica' has pink flowers and is 1 foot.

'Heidekind' has many short spikes of deep rose pink, 8 to 10 inches.
'Rosea'has rose pink flowers, 18 inches.
'Rubra' has deep rosy red flowers.

Hybrids:

'Goodness Grows' (*V. alpina 'Alba' x V. spicata*) is a long-flowering blue with a low-growing habit, 10 to 12 inches.

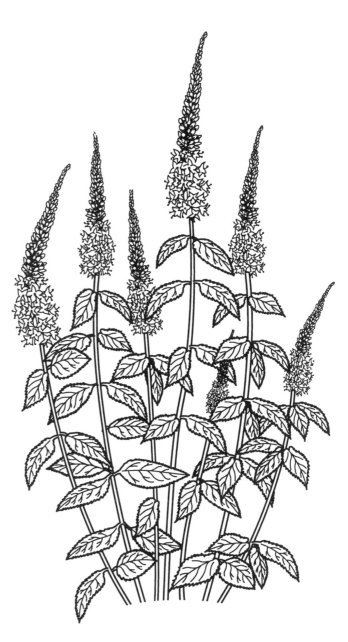

'Icicle' (probably *V. subsessilis x V. spicata*) is an attractive, long-flowering white veronica and 2 feet tall.

'Noah Williams' is a white-flowered, variegated speedwell, 1 foot.

'Sunny Border Blue' (*V. spicata x V. longifolia*) is a long-flowering, dark blue with attractive, toothed, puckered, glossy leaves, 2 feet. 1993 Perennial Plant of the Year

Problems: None serious.

Propagation: Division and early summer cuttings. Seed (variable).

Zones: 3-9

Related Species:

V. austriaca ssp. *teucrium*, Hungarian Speedwell, is 1- to 2-feet edging (foreground) plant with a mounded sprawling habit compared to the more erect habits of the others. Peastaking is advised. Leaves are crenate to serrate.

'Blue Fountain' has bright blue flowers and is 2 feet tall.

'Crater Lake Blue' is vivid blue, mounded (to sprawling), and a good 10- to 15-inch edging plant.

'Kapitan' has blue flowers and is 10 inches.

'Royal Blue' is blue and 12 to 18 inches.

'Shirley Blue' is bright blue and 10 inches.

V. gentianoides, Gentian Speedwell, forms a basal rosette of green entire to slightly crenate leaves, flowers earlier than the other speedwells, and is 1 to 2 feet tall.

'Alba' has white flowers, tinged blue.

'Pallida' has light blue flowers, 12 to 15 inches.

'Variegata' has pale blue flowers with creamy white leaf variegation, 14 to 18 inches.

V. incana, Woolly Speedwell, is similar to *V. spicata* except for grayish white tomentose pubescence on the leaves. The leaves have crenate margins.

'Baccarole' has rose pink flowers, gray green leaves, and is 12 to 18 inches tall.

'Minuet' has bright pink flowers, silvery green foliage, and is 12 to 15 inches.

'Nana' has white flowers, velvety leaves with a spreading habit, 8 to 10 inches.

'Romilley Purple' has a deep lavender blue flower and 2-foot bushy habit.

'Saraband' has lavender blue flowers, gray-green leaves, and grows 1 to 1½ feet.

'Silver Carpet' has large silvery gray leaves and dark blue flowers.

'Wendy' is lavender blue with gray foliage, 1½ to 2 feet.

V. longifolia, Long-leaf Speedwell, are generally taller (2 to 4 feet) with longer green leaves with doubly serrate margins. Most plants are sturdy and do not need staking.

'Alba' is white flowered and 2 to 3 feet.

'Blue Giant' is lavender blue and 3 to 4 feet.

'Forster's Blue' has long-lasting dark blue flowers and is 2 feet tall.

'Snow Giant' is 2 to 3 feet and white flowering.

V. subsessilis has lavender blue racemes on a multi-branched, 1½- to 2½-foot plant.

Other Late Summer to Fall Perennials

Acanthus spinosus — Spiny Bear's Breeches

Partial shade, well-drained site.

Unique spiny purple or white spikes.

Showy pinnately parted bold foliage.

Upright mounded, specimen, 3 to 4 feet.

Alcea rosea — Hollyhock

Full sun, well drained. Singles or doubles in every color (not blue). 'Nigra' dark maroon, appears black.

Showy old-fashioned colorful racemes.

Large rugose leaves.

Upright, 4 to 8 feet, background.

Anaphalis triplinervis

12 to 18 inches.

Anaphalis margaritacea

3 feet.

Pearly Everlasting

Full sun, partial shade, well drained.

Miniature white "strawflowers" in corymbs, good fresh or dried flowers.

Woolly, grayish-white, tomentose leaves.

Mounded, foreground to midborder.

(Larva always destroys my plants.)

Chelone lyonii — Turtlehead

Full sun, partial shade, moist site.

Unique pink flowers on racemes.

Upright, 3 feet, midborder, background.

Cimicifuga racemose — Black Snakeroot

4 to 6 feet.

C. simplex — Kamchatka Bugbane

3 to 4 feet.

Partial shade, moist site.

White showy racemes, compound foliage, two tiered, background.

Dendranthema zawadskii (rubellum)

'Clara Curtis' pink, yellow centers, 2 to 3 feet.

'Duchess of Edinburgh' faded red, 2 feet.

Full sun, well drained.

Single daisy flowers, dissected pubescent leaves.

Mounded, midborder.

Hibiscus moscheutos — Hibiscus, Rose Mallow

Full sun, well drained, moist.

Large 5- to 12-inch solitary flowers, white, pink, red, bicolors, single.

Large, broad, orate, pubescent leaves.

Mounded, 4 to 5 feet, background.

Kirengoshoma palmata — Yellow Waxbells

Partial shade, moist, acidic site.

Yellow nodding bells.

Handsome large maple-like leaves with dark purple stems.

Mounded, 3 to 4 feet, specimen, midborder to background.

Ligularia dentata — Bigleaf Goldenray

Dappled shade, moist site.

Orangish yellow daisy, 3 to 4 feet.

Large round leaves.

Mounded, midborder to background.

L. 'The Rocket' — Rocket Ligularia

Dappled shade, moist site.

Large yellow racemes, 5 to 6 feet.

Large toothed leaves.

Upright mounded, background.

Liniope spicata — Creeping Lily

Turf, ground cover.

Partial shade.

Thin lavender racemes, blackberry-like fruit in fall.

Dark green arching grass-like foliage

Evergreen.

Macleaya cordata — Plume Poppy

Full sun, well drained, spreads.

Large fluffy panicles.

Large attractive lobed leaves.

Upright, 6 to 8 feet, background.

Physalis alkekengi — Chinese Lantern

Grown for inflated orange calyx, resembling a lantern.

Spreads, confine or containerize, 1 to 2 feet.

Solidago cultivars — Goldenrod

Full sun, well drained.

Yellow fluffy curving panicles.

Mounded, 2 to 3 feet, midborder.

'Golden Fleece' is multi-branched and 18 inches tall with lots of bright yellow late season color.

Thalictrum rochebrunianum

Lavender Meadow Rue

Partial shade, sun, moist, well drained.

Lavender showy stamens, no petals.

Foliage resembles maidenhair fern.

Upright, two tiered, 4 to 6 feet. Background.

T. delavayi 'Hewitt's Double' — Yunnan Meadow Rue

Partial shade, moist, well drained.

Double purple flowers with maidenhair fern-like foliage.

Upright, 4 feet, background.

Tricyrtis hirta — Toad Lily

Partial shade, moist.

Unique pale lavender flowers with dark purple spots.

Veined alternate foliage.

Mounded, 2 to 3 feet, midborder.

MID TO LATE SUMMER BULBS

Lycoris squamigera

Magic Lily, Surprise Lily, Resurrection Lily, Naked Lady
Hardy or Autumn Amaryllis
Amaryllidaceae (Amaryllis Family)
○, ◑ Foreground, Midborder, Late summer

Flowers: Numerous pink or rose lavender flowers are borne in an umbel on a leafless stem in late summer. Each flower is trumpet shaped, 3 to 4 inches long, and fragrant. The flower parts (pistil and filaments) are pink and quite showy.

Leaves: The 12-inch leaves appear in spring. They are long, flat, and rounded at the tip. They die back in late June. These leaves are similar to *Narcissus* but are slightly wider and greener.

Growing Needs: Plant this bulb in midsummer in a partial shade to full sun location. The soil should be well drained. This plant needs a dry dormant period before flowering. The bulb also resembles *Narcissus* but has an elongated neck, which makes it easy to distinguish from daffodils when digging them up in the garden.

Landscape Use:

Height:	Foliage—15 inches	
	Flowers—18 inches to	
	2 feet	
Planting depth:	6 inches	
Planting time:	Summer	
Spacing:	6 inches	
Texture:	Medium	

Use in masses of 10 to 12 interplanted in plants with attractive summer foliage. It adds some zip to the late summer garden. Surprise lilies are good cut flowers.

Design Ideas: Crater Lake Blue veronica and miniature roses provide a good foliage foundation in front of the leafless surprise lily stems. Magic lily is attractive in front of dwarf shrubs, like dwarf cranberry bush, with *Coreopsis rosea* (repeats its pink flower color) and catmint as edging plants in front of it. Combine with astilbe and Jacob's ladder foliage in a partially shady spot.

Problems: None serious.

Propagation: Summer division.

Zones: 5-9

Other Late Summer to Fall Bulbs

Colchicum —Autumn Crocus

6 to 10 inches. Full sun, light shade, well-drained site.

Large pink, lilac, white flowers (with showy yellow flower parts) appear without foliage.

Plant in late summer 3 inches deep.

Naturalized in lawn or woodland edge.

C. autumnale —

3 to 6 inches, earliest, late summer, pink to white, 4 to 6 inches wide.

'Alboplenum' — double white flowers.

'Album' — white flowers.

'Plemun' — double lilac flowers.

C. bornmuelleri —

6 inches, late summer, rose purple, white throat, with purple-brown anthers.

C. byzantium —

6 inches, early fall, lilac to pink, 3 to 4 inches across.

C. cilicicum —

5 inches, early fall, deeper rose lilac, star-shaped.

C. speciosum —

5 inches, fall, shades of rose-purple, 5 to 6 inches across.

Colchicum Hybrids

'Autumn Queen' lilac, slightly checkered, 6 inches tall, fall.

'Giant' large-flowered, violet with white throat, 8 inches tall, early fall.

'Lilac Wonder' large lilac with white center lines, late fall, 8 inches tall, late fall.

'Violet Queen' large purple checkered with white throat, late summer, 7 inches.

'Waterlily' large double, deep rose-pink, 6 inches tall, mid to late fall.

Crocus speciosus — Fall Crocus

2 to 4 inches. Full sun, well drained, easy to grow. Lavender blue flowers before the leaves, early to mid-fall.

'Albus,' white.

'Artabir,' lavender with deeper veins, creamy throat.

'Cassiope' large pale violet, yellow throat, late fall.

A sampler of other hardy fall crocus:

C. kotschyanus (zonatus)—4 inches, early fall. Pale lilac, darker veins, yellow throat.

C. ochroleucus — 3 inches, late fall. Creamy white, yellow throat.

C. pulchellus — 4 inches, fall. Lilac, veined, white anthers.

# Glossary

alternate: The arrangement of plant parts, such as single leaves positioned one at a node on different sides along the stem.

background plant: A plant that serves as a backdrop for smaller plants; usually 3 feet and taller.

basal: Pertaining to the base of the plant.

basal rosette: Leaves all emerge from a single point (or rosette) out of the soil.

biternate: When each division of a ternate leaf is also ternate.

bract: A modified leaf associated with the floral axis or subtending a flower.

bulb: A modified underground stem surrounded by scale-like leaves. The bulb contains stored food for the undeveloped shoots of the new plant.

calyx: The outer protective covering of a flower, usually leaf-like but may also be petal-like.

compound leaf: A leaf composed of two or more leaflets on a common petiole.

cordate: Heart-shaped.

corm: A modified underground stem that lacks scales, i.e., crocus.

corolla: Petals of a flower.

corymbose: A flat-topped, indeterminate inflorescence in which the flower stalks grow upward at different points to approximately the same height, the outer flowers open first.

crenate: Round or scalloped toothing on the leaf margin.

cultivar: A cultivated variety. A plant that is different in a distinguishable way from the species in either its appearance, growth characteristics or make-up and can be reproduced "true to type" either sexually or asexually, retaining the desired characteristic.

cyme: A more or less flat-topped determinate inflorescence that blooms from the center of the flower cluster out to the edges.

deadhead: To remove spent flowers.

disk flower: A flower with a tubular corolla on the head of a plant belonging to Asteraceae.

edging plant: A plant that tends to mound rather than spread; usually 6 to 12 inches.

elliptic: Wider at the center tapering to equal tips.

falls: The outer whorl of tepals of the Iris flower, which are positioned downward or are "falling down".

forcing: Growing plants within a greenhouse in order to make them flower before they would naturally.

foreground plant: A plant placed just behind the edging plants; usually 1 to 2 feet in height.

glabrous: Lacking hairs, smooth.

glaucous: Covered with a waxy coating.

ground cover plant: A plant that spreads laterally and covers the ground; height is usually not more than 12 inches tall.

habit: The general appearance or shape of a plant in the landscape.

Examples —

> *clumped*: A clustered mass.
>
> *creeping* (sprawling): Growing along the surface.
>
> *mounded* (low, upright): Being rounded on the top and straight on the sides.
>
> *prostrate*: Growing flat along the ground.
>
> *two-tiered*: The flowers are borne above the foliage giving the flowers the illusion of floating over the leaves.
>
> *vase-shaped*: Plant habit resembles a vase, wider at the top than the base.
>
> *upright*: Erect or tall plant habit with little rounding or mounded appearance. Fine-textured or more open variations include upright airy or upright lollipop.

hispid: Covered with short, rigid hairs.

inflorescence: Producing blossoms; flowering.

interplanting: Planting plants within established plants.

lanceolate: A leaf that is lance-shaped, tapering at each end.

linear: Long and narrow, with parallel or nearly parallel sides.

midborder plant: A plant placed between the foreground and background plants; usually 2 to 3 feet tall.

midvein: The main or central vein of a leaf (midrib).

mottled: A spot or blotch of color.

mulch: A covering of various substances to protect from evaporation, freezing, and to control weeds.

naturalize: To allow a plant to grow as in nature.

oblanceolate: A leaf that is broader at the apex and tapering at the base.

oblong: A leaf that is much longer than wide and is parallel at the sides.

obovate: A leaf that is egg-shaped with the narrow end at the base.

opposite: The arrangement of leaves in pairs along the stem at different heights.

ovate: A leaf that is egg-shaped with the wider end at the base.

palmate: The arrangement of leaflets or lobes arranged radially from a common point.

panicle: An indeterminate inflorescence that is loosely and irregularly branched.

peduncle: The main stalk of an inflorescence.

petaloid: Resembling a petal.

petiole: The stalk by which a leaf is attached to a stem.

pinnate: A compound leaf in which the leaflets are arranged in a feather-like fashion.

polymorphic: The occurrence of many different forms on the same plant.

propagation: A means to reproduce; by seed, division, or cuttings.

pubescent: A covering of soft, short hairs.

raceme: An unbranched indeterminate inflorescence in which stalked flowers are arranged singly along a stem.

ray flower: An asymmetrical flower with a strap-like corolla, at the margin of the head of a plant belonging to Asteraceae.

rosette: A circular cluster of leaves.

rhizome: A root-like horizontal stem growing under or along the ground that sends roots from the lower surface and shoots from the upper.

salverform: Funnel-shaped.

scape: A leafless flower stalk.

sepal: One of the segments forming the calyx.

serrate: An edge or margin of a leaf that is notched.

sessile: Without a stalk, attached directly at the base.

spatulate: A leaf that is shaped like a spatula or spoon.

spike: An elongated inflorescence with stalkless flowers.

spur: A tubular extension of the corolla or calyx of a flower.

stamen: The organ of a seed plant that bears pollen.

standards: The inner whorl of tepals of the Iris flower.

stellate: Star-shaped, or radiating from a center.

stipule: A leaf-like appendage at the base of a leaf.

succulent: Thickened, fleshy plant tissues.

ternate: Having three leaflets.

texture: The surface appearance.

tomentose: Covered with dense, short hairs.

trifoliate: With three leaflets.

tuber: A swollen underground stem bearing buds from which new shoots arise.

umbel: A flat to rounded determinate inflorescence in which the flower stalks arise from a common point.

variegated: Having streaks, marks, or patches of different color(s).

whorled: The arrangement of three or more leaves or buds at a node.

zone: An area distinguished by a distinctive feature or character (i.e., USDA Hardiness Zones distinguished by temperature).

Bibliography

The American Garden Guides. *Perennial Gardening*. New York: Knopf Publishing Group, 1994.

The American Horticultural Society. *Encyclopedia of Gardening*. New York: Dorling Kindersley, 1993.

Bailey, Liberty Hyde. *Hortus Third*. New York: Macmillan Publishing Co., 1978.

Biondo, Ronald J. and Charles B. Schroeder. *Introduction to Landscaping: Design, Construction, and Maintenance,* Second Edition. Danville, IL: Interstate Publishers, Inc., 2000.

Bird, Richard. *The Cultivation of Hardy Perennials*. London: B.T. Batsford Ltd., 1994

Bryan, John E. *Bulbs*. New York: Macmillan Publishing Co., 1994.

Clausen, Ruth Rogers and Nicholas H. Ekstrom. *Perennials for American Gardens*. New York: Random House, 1989.

Cox, Jeff and Marilyn Cox. *The Perennial Garden*. Emmaus, PA: Rodale Press, 1985.

Druse, Ken and Margaret Roach. *The Natural Habitat Garden*. New York: Clarkson N. Potter, Inc., 1994.

Giles, F.A., Betty R. Hamilton and T.B. Voigt. *Ground Covers for the Midwest*. University of Illinois at Urbana-Champaign. 1983.

Giles, F.A., Rebecca M. Keith and Donald E. Saupe. *Herbaceous Perennials*. Reston, VA: Reston Publishing Co., 1980.

Griffths, Mark. *The New Royal Horticultural Society Dictionary: Index of Garden Plants*. Portland, Oregon: Timber Press, 1994.

Harper, Pamela J. *Designing with Perennials*. New York: Macmillan Publishing Co., 1991.

Huxley, Anthony, Editor-in-Chief, Mark Griffiths, Editor, and Margot Levy, Managing Editor. *The New Royal Horticultural Society Dictionary of Gardening* in 4 volumes. New York, NY: The Stockton Press. 1992.

Phillips, Ellen and Colston Burrell. *Rodale's Illustrated Encyclopedia of Perennials.* Emmaus, PA: Rodale Press, 1993.

Phillips, Roger and Martyn Rix. *Perfect Plants.* New York: Random House, 1996.

Pirone, Pascal P. *Diseases and Pests of Ornamental Plants.* 5th ed. New York: Wiley-Interscience Publishers, 1978.

Schroeder, Charles, B., Eddie Dean Seagle, Lorie M. Felton, John M. Ruter, William Terry Kelley, and Gerard Krewer. *Introduction to Horticulture*, Third Edition. Danville, IL: Interstate Publishers, Inc., 2000.

Scott-James, Anne. *Perfect Plant, Perfect Garden.* New York, NY: Summit Books, 1988.

Sheldon, Elisabeth, *A Proper Garden: on Perennials in the Border.* Harrisburg, PA: Stackpole Books, 1989.

Springer, Lauren. *The Undaunted Garden.* Golden, Colorado: Fulcrum Publishing, 1994.

Still, Steven M. *Manual of Herbaceous Ornamental Plants. 3rd ed.* Champaign, IL: Stipes Publishing co., 1994.

Taylors Guide Staff; *Taylors Guide to Perennials.* Rev. ed. Boston: Houghton Mifflin Co., 1986.

Thomas, Graham Stuart. *Perennial Garden Plants or The Modern Florilegium.* Portland, Oregon: Sagapress, Inc./ Timber Press, Inc., 1994.

Time Life Staff. *The BIG Book of Flower Gardening.* Alexandria, VA: Time Life Books, 1996.

Zomlefer, Wendy B. *Guide to Flowering Plant Families.* Chapel Hill, NC: The University of North Carolina Press, 1994.